The American Foundrymen's Society

CASTING BUYER'S GUIDE

For buyers, designers, engineers, specifiers, producers, and users of castings

First Edition

ISBN 0-87433-107-2

First Printing, April 1988
Second Printing, October 1988
Third Printing, January 1998
Fourth Printing, December 1999

Printed in the U.S.A.

American Foundrymen's Society, Inc.
Des Plaines, Illinois 60016-2277

Table of Contents

Author's Preface

Castings are widely used in all types of applications due to exceptional characteristics and the great design flexibility inherent in a variety of casting processes.

Individual foundries have long used their own standards covering various aspects of producing castings. In many cases these may vary widely from one supplier to another. In addition, however, newer techniques and molding processes also contribute to variations. The variations in existing information, along with the lack of data in many areas, indicated the need for a guide for persons involved in procuring castings and with which they could expect foundries to base proposals and produce castings to a specific quality level.

The American Foundrymen's Society, as a service to the casting customer of the industry, has undertaken the preparation of this guide for casting buyers and foundry production personnel as to the most economical casting processes.

The information provided in this guide is not intended to be limiting in any way, but rather is expected to provide consistency in terminology and definition resulting in a better understanding between supplier and user.

Norris Luther

Introduction

Casting procurement is one of the most difficult of all industrial purchasing activities. In recent years efforts have been made to bring together buyers and producers of castings in the hope of promoting a better understanding of their mutual problems.

It is important that users be aware of the range of casting production processes available, the relative costs of both tooling and castings, and the various parameters which affect the buying decision.

Castings can be produced with a variety of molding processes, surface finishes, dimensional accuracies, finish allowance, and drafts. The flexibility of the various processes for producing castings give the engineer and manufacturer wide latitude in design to meet specific requirements.

This guide is for the individual involved in purchasing a usable, unmachined casting consistent with normal production practices, reproducibility, reasonable mold or pattern life and maintenance costs, normal inspection, packing, and shipping procedures. Special requirements for finish, tolerances, etc., beyond the standard produced within a specified casting process may be specified where required, although additional costs may be involved. Consultation with the foundry will usually result in such requirements being properly considered in the quotation and mutually understood. Conversely, more liberal values should be indicated where acceptable since they tend to keep costs to a minimum.

These suggested guidelines are not designed to reduce quality level, lessen competition, restrict customer requirements, or increase costs. On the contrary, knowledge of the various casting processes will improve understanding of quality levels, widen the area of available sources of supply, increase the customer's ability to distinguish special requirements, and reduce costs by eliminating the specification of unnecessary rigid limits where they are not required.

The aim of this book is therefore to provide a guide to good practice for the purchase of castings—to assist designers and specifiers, buyers and users of castings—by outlining some of the more important considerations necessary for effective, comparatively trouble-free procurement. It is not intended as a work of technical reference, although of necessity some technical information has been included.

1

Metal Casting and Molding Processes

The casting process consists of pouring molten metal into a mold containing a cavity of the desired shape of the casting. Metal casting processes can be classified either by the type of mold and pattern, or by the pressure or force used to fill the mold with molten metal. Conventional sand, shell, and other chemically bonded molding processes utilize a permanent pattern, but the mold is used only once. Permanent molds and die-casting dies are machined in metal, graphite, or other specifically selected die matrix materials and are used for large volume production castings. Investment casting and the expanded polystyrene molding processes involve an expendable mold as well as an expendable pattern for each casting produced.

Each process must consider the variables which are present as the molten metal meets the mold face. The evolvement of the decomposition by-products of the additives and binders used can affect the final cast surfaces, the casting detail, and dimensional accuracy in all metals.

1.1 FACTORS AFFECTING PROCESS SELECTION

Some of the important factors affecting the selection of a casting process are as follows:

- Quantity of castings required
- Design of the casting
- Tolerances required
- Complexity
- Metal specification
- Surface finish required
- Tooling costs
- The economics of machining versus casting costs
- Financial limits on capital costs
- Delivery requirements

More than one process may be possible and the distance or accessibility to a supplier's foundry may be important. Usually it is possible to devise a simple formula to assist in the ultimate decision using such data as the capital cost of equipment, the unit cost of the casting multiplied by the volume, and making allowances for special factors.

1.2 GENERAL DESCRIPTION OF BASIC PROCESSES

The casting buyer may not be familiar with the casting terminologies, dimensional tolerances, advantages, and disadvantages associated with the various casting processes noted below:

Conventional Molding Processes

- Green Sand Molding
- High-Density Molding (High Squeeze Pressures, Impact)
- Flaskless Molding
- Tight Flask Molding
- Skin-Dried and Dry Sand Molding

Precision Molding and Casting Processes

- Permanent Molding ("Gravity Die Casting")
- Low-Pressure Molding (Die Casting)
- High-Pressure Molding (Die Casting)
- Investment Casting (Lost Wax)
- Ceramic Molding (Shaw Process)
- Hitchiner Process (CLA, CLAS, CLAV)

Chemically Bonded Sand Molding Processes

- Shell Molding (Organic)
- Sodium Silicate CO_2 Bonded Molding (Inorganic)
- No-Bake Molding (Chemically Bonded Self-Setting Sand Mixtures)(Organic)

Special and Innovative Molding and Casting Processes

- Evaporative Pattern Casting (EPC)
- Vacuum ("V") Process Molding
- Centrifugal Process Molding
- "H" Process Molding

This section provides a brief description of the major molding processes used to produce castings. During the past two decades, innovative molding and casting techniques have been developed and implemented. This section describes only those which are used in the foundry industry today. Should you have questions concerning a molding or casting process not described, it is suggested that you contact the American Foundrymen's Society (AFS Library).

1.3 CONVENTIONAL MOLDING PROCESSES

This section describes the conventional molding processes which are the most widely used in the foundry industry.

1.3.1 Green Sand Molding

Green sand is by far the most diversified molding method used in present casting practices. The process utilizes a mold made of compressed or compacted moist sand. The term "green" denotes the presence of moisture in the molding sand, and indicates that the mold is not baked or dried. The mold material consists of silica sand mixed with a suitable bonding agent (clay) and moisture. To produce the mold a flask, usually a metal frame, although wood may be used for some processes and types of castings, is placed over the pattern to produce a cavity representing one half of the casting. Compaction is achieved by either jolting or squeezing the mold. The other half of the mold is produced in like manner and the two flasks are positioned together to form the complete mold.

If the casting has hollow sections, a core consisting of hardened sand (baked or chemically hardened) is used. Cores are located in pockets formed by projections on the pattern equipment to produce coreprints. Should extra support for the core be required, chaplets or spacers are properly positioned to maintain the required dimension. These will fuse with the molten metal when the casting is poured. Green sand is the best known of all the sand casting molding methods, as the molds may be poured without further conditioning. This type of molding is most adaptable to light, bench molding for medium-sized castings or for use with production molding machines.

- *Advantages*

  Most metals can be cast by this method. Pattern costs and material costs are relatively low. The method is adaptable to large or small quantities.

- *Disadvantages*

  There are practical limits to complexity of design. Machining is often required to achieve the finished product. Dimensional accuracy cannot be controlled as well as with other molding processes, although good standards are possible with quality pattern equipment, modern process controls, and high-density molding.

1.3.2 High-Density Molding (High Squeeze Pressure, Impact)

In recent years the introduction of high-pressure molding techniques has greatly improved the standards of accuracy and finish which can be achieved with certain types of castings.

Dramatic changes have occurred in the foundry molding system by increasing the capacity and ability of the molding machines to produce molds. This has been accomplished by more effective use of large air cylinders, application of electronics, use of hydraulics, and the innovation of an explosive-type method of compacting the sand around the pattern, referred to as impact molding.

In recent years high-pressure or high-density molding has been widely adapted for the casting of most metals. High-pressure (high-density) molding has virtually eliminated the past problems of mold-wall movement. Mold-wall movement has been markedly aggravated by the moisture content of the molding sand, the density of the molds, and the mold surface hardness.

High-pressure molding practices have allowed lower moisture contents in the molding sand, so that higher mold densities can be achieved. Thus, castings having considerably better dimensional accuracy are produced, the surface finish of the casting is improved, and they are closer to pattern accuracy.

- *Advantages*

  Most metals can be cast by this process. Castings with closer dimensional tolerances, consistently uniform casting weight, better surface finish, increased productivity, and lower cost. Other advantages are reduced feed metal, improved casting soundness, reduced cleaning costs, and minimum setup for machining because of casting dimensional accuracy.

- *Disadvantages*

  Tighter quality controls are required and as a result, the foundry using high-pressure (high-density) molding is generally accustomed to more sophistication in equipment, maintenance, and operating procedures.

1.3.3 Flaskless Molding

One recent innovation in green sand molding has been the introduction of flaskless molding — with both vertical as well as horizontal partings. Contrary to any misconceptions, a flask must be used on all green sand molding primarily for containment of sand while it is compacted about the pattern. In flaskless molding (whether vertical or horizontal) instead of using "tight" individual flasks for each mold produced, the master flask is contained as an integral unit of the totally mechanized mold producing system. Once the mold has been stripped from the integral mold producing unit, it is held against the other half of the mold with enough pressure to allow pouring of the metal.

Through advanced engineering techniques as well as constant modification and improvements, vertical flaskless molding has achieved notable production as well as casting quality and has attained new heights of casting dimensional tolerance and accuracy. The vertical flaskless systems is suited to gray, malleable, and ductile iron as well as steel, aluminum, and brass. In the vertical flaskless system, the completely contained molding unit blows and squeezes a mold against a pattern (or multiple patterns) which has been designed for a vertical gating system. Molds of this type can be produced in very high quantities per hour, and of high density (mold hardness ranging from 85-95 B scale) with excellent dimensional reproducibility.

- *Advantages*
  No expenditure required for flasks. No cleaning or maintenance of flasks. Improved working conditions: no handling, storing, or shakeout of flasks.

- *Disadvantages*
  Restrictions apply to size of casting, use of complicated cores and core assemblies, and number of castings per mold. Mold handling may be more difficult.

1.3.4 Tight Flask Molding

The tight flask molding process uses a flask (made of metal or wood) to contain the sand during the mold producing operation. The flask remains around the sand during the core setting, closing of the mold, pouring and cooling of the casting. After cooling, castings and sand are separated from the flasks, which are then reused to produce more molds.

- *Advantages*
  More flexibility in type and number of castings poured per mold. Increased utilization of total flask surface when compared to flaskless moldings. Increases the size and weight of a casting which may be poured. Use of complicated cores or core assemblies is not a problem compared to flaskless molding.

- *Disadvantages*
  Flask handling systems require capital investment and a good maintenance program. The variety and size of castings produced in a "tight" flask molding line may require changing the flask for a larger size casting.

1.3.5 Skin-Dried and Dry Sand Molding

The process of producing molds with different surface characteristics from a green sand material is a common practice and is noted below for two different types of molds.

Skin-dried or air-dried molds are sometimes preferred to green sand molds where assurance is desired that the surface moisture and other gas-forming materials are lowered. By skin drying the face of the mold after special bonding materials have been added to the sand molding mixture, a firm mold face is produced similar to that obtained in dry sand practice. Shakeout of the mold is almost as good as that obtained with green sand molding. Skin-dried molds are commonly employed in making medium-heavy and heavy castings. Generally the surface of the mold is washed or sprayed with a refractory mold coating. The most common method of drying the refractory mold coating uses hot air, gas, or oil flame.

Skin drying of the mold can be accomplished with the aid of torches, a bank of radiant heating lamps, or electrical heating elements directed at the mold surface.

- *Advantages*
  Reduces surface moisture and other gas-forming materials from mold. Commonly used in the production of medium-heavy to heavy castings.

- *Disadvantages*
  Is more expensive to produce. Mold sections must be completely dry and cool prior to assembly.

Dry sand molding is the green sand practice modified by baking the mold at 400F to 600F (204C to 316C). Some foundries use dry sand molds to produce intricate parts which are difficult to cast to exact size and dimensions. Molds are generally dried (or baked) in large mold drying ovens, or with large mold heaters. Castings of large or medium size, and of complex configuration such as frames, engine cylinders, rolls, large gears and housings are often made utilizing the dry sand technique. Both ferrous and nonferrous metals are cast in this type of mold.

- *Advantages*
  Dry sand molds are generally stronger than green sand molds and therefore can withstand much additional handling. The castings usually possess more exactness of dimension than if they were molded in green sand. The improved quality of the sand mixture due to the removal of moisture can result in a much smoother finish on the castings than if made in green sand molds. Where molds are properly washed and sprayed with refractory coatings the casting finish is further improved.

- *Disadvantages*
  This type of molding is much more expensive than green sand molding and is not a high-production process. Correct baking (drying) times are essential.

1.3.6 Sand-Mold Casting Tolerances

The sand-mold casting tolerances may vary considerably depending upon the type of metal used. For a comparison of tolerances please refer to the table at the top of page 7.

Sand-Mold Casting Tolerances

Metal	Tolerance, ± inch/foot*
Aluminum Alloys	1/32
Cast Iron	3/64
Copper-Base Alloys	3/32
Cast Steels	1/16
Magnesium Alloys	1/32
Malleable Iron	1/32

* *Refer to parting line influence, page 18.*

1.4 PRECISION MOLDING AND CASTING PROCESSES

This section describes molding processes which produce castings with an improved finish, provides excellent detail, with a higher degree of dimensional accuracy.

1.4.1 Permanent Mold

The foundry term "permanent mold" is used to describe a mold that can be used repeatedly. In contrast to the sand mold which must be destroyed to remove the casting, the permanent mold is designed so that it can be separated to eject and remove the solidified casting. In the current practice, the molds are usually made of cast iron or steel, although graphite, copper, and aluminum have been used as mold materials. Permanent mold castings can be produced from all of the metals including iron and copper alloys, but are usually light metals such as zinc-base, magnesium, and aluminum.

Permanent Mold Casting of Ferrous Alloys

Because of the higher pouring temperature, the casting of iron in permanent molds developed much more slowly than for nonferrous metals. The technique of using permanent or reusable molds for ferrous metal castings to produce small ferritic iron castings for the automotive and appliance industries was first undertaken about 1925. After a development period of several years, the process was commercially established. The permanent mold casting of soil pipe fittings was introduced in 1955 and more recently it has been applied to the production of gray iron pressure pipe fittings. Current production practices are designed almost entirely for small ferritic gray and ductile iron castings.

1.4.2 Gravity Permanent Mold

The flow of metal into a permanent mold under gravity conditions is referred to as a gravity permanent mold. There are two techniques in use: static pouring, where metal is introduced to the top of the mold through downsprues similar to sand casting, and tilt pouring, where metal is poured into a basin while the mold is in a horizontal position and flows into the cavity as the mold is gradually tilted to a vertical position.

Normally, gravity molding is used because it is more accurate than shell molding. It is preferred almost exclusively to shell molding for light alloy components.

- *Advantages*

Produces a sound dense casting with superior mechanical properties. Since the mold is normally made of metal and is relatively stable, the castings produced are quite uniform in shape and often have a higher degree of dimensional accuracy than castings produced in sand, which reduces or eliminates some of the machining that might be required on the part. The permanent mold process is also capable of producing a consistent quality of finish on the castings. The process also lends itself very well to the use of expendable cores and makes possible the production of parts that for one reason or another are not suitable for the pressure die-casting process.

- *Disadvantages*

The cost of tooling is usually higher than for sand castings and must be amortized over a large number of castings. The process is generally limited to the production of somewhat small castings of simple exterior design, although complex castings such as aluminum engine blocks and heads are now commonplace.

1.4.3 Low-Pressure Permanent Mold

Low-pressure permanent mold is exactly what its name implies, namely a method of producing a casting by using a minimal amount of pressure (usually 5 to 15 pounds per square inch) to fill the die. It is a casting process that helps to further bridge the gap between sand and pressure die casting. It lends itself well, but it is not necessarily limited to aluminum castings with wall thicknesses of 5/32 inch or heavier. Low pressure compares favorably with pressure die casting and process capability, but has a slower cycle and therefore it is better suited to lower quantity requirements. It should therefore be classified as a supplementary process for producing castings economically. Many castings now produced by the sand casting or gravity permanent molding process can also be cast by the low-pressure process.

- *Advantages*

Good dimensional accuracy. Consistent quality. Relatively inexpensive castings. Suitable for fairly complex castings in light alloys, such as cast aluminum wheels.

■ *Disadvantages*
Relatively high cost of equipment. Commonly used for light alloy castings. Other metals can be cast using the low-pressure process but few foundries produce metals other than light alloy.

1.4.4 High-Pressure Die Casting

The high-pressure die-casting process is widely used for producing large volumes of zinc, aluminum, and magnesium castings of intricate shape. The essential feature of die casting is the use of permanent metal dies into which the molten metal is injected under high pressure (normally 5,000 psi or more).

The rate of production of die casting depends largely on the complexity of design, the sectional thickness of the casting, and the properties of the cast metal. Great care must be taken with the design and gating to avoid high-pressure porosity to which this process is prone. Where pressure tightness is essential the buyer can arrange for special processing such as "vacuum die casting."

■ *Advantages*
Low cost of castings, high degree of complexity and good surface finish. Considerable design flexibility. High degree of accuracy. As with other processes inserts can be cast in if needed to enhance design capability. Suitable for relatively low melting point metals (1,600F) (871C) (lead, zinc, aluminum, magnesium, and some copper alloys) and high volume.

■ *Disadvantages*
Limit on size of castings. Most suitable for small castings up to approximately 75 pounds. Equipment and die costs are high. Some risk of porosity. Good design is essential.

1.4.5 Investment Casting (Lost Wax)

Investment denotes the mechanical manner of obtaining a mold rather than the material used. It is the process of completely investing a three-dimensional pattern in all of its dimensions to produce a one-piece destructible mold into which molten metal will be poured.

The principal feature of investment molding is that a fluid mixture flows around the wax pattern providing excellent detail. This mixture sets with the aid of a hydraulic bond. These wax patterns are assembled on a "tree" and invested with a ceramic slurry. The tree is then immersed in a fluidized bed of refractory particles to form the first layer of the ceramic shell. The mold is allowed to dry and the process repeated with coarser material until sufficient thickness has been built up to withstand the impact of hot metal. The wax is then melted out for subsequent recovery and the molds are fired prior to casting. Most materials can be molded by this process but the economics indicate that fairly high volume is necessary and the shape and complexity of the castings should be such that savings are made by eliminating machining. Unless advantage can be taken of these features, it is unlikely that investment casting will compare favorably with other processes. Accuracy of castings is totally dependent on the accuracy of the die and extremely fine tolerances can be achieved with an exceptionally wide range of materials.

- *Advantages*
  Extreme accuracy and flexibility of design. Useful for casting alloys that are difficult to machine. Exceptionally fine finish. Suitable for large or small quantities.

- *Disadvantages*
  Limitation on size of casting. Casting costs make it important to take full advantage of the process to eliminate all machining operations.

1.4.6 Ceramic Molding (Shaw Process)

Ceramic molding can be accomplished through two diverse techniques:

1. True ceramic molding.
2. Ethyl silicate slurry molding (also known as Shaw process, Avnet-Shaw Osborn-Shaw process, and the Dean process).

Ceramics are materials which are made from a clay base and contain various oxides and ingredients other than sand. The raw clays are calcined or fired at high temperatures and are then blended, mixed with water, formed into mold components, and then fired.

In true ceramic molding, the refractory grain can be bonded with calcium or ammonium phosphates. The preferred methods for producing ceramic molds is the dry pressing method in which molds are made by pressing the clay mixture containing 4 to 9 percent moisture in dies under a pressure of 1 to 10 tons per square inch. After pressing, molds are stripped from the dies and then fired at temperatures between 1,650F and 2,400F (899C and 1,316C).

The ethyl silicate variation is accomplished in the following manner: a mixture of a graded refractory filler, hydrolyzed ethyl silicate, and a liquid catalyst are blended together to form a slurry consistency. The slurry is then poured over a pattern and allowed to jell. After gelation, the mold is stripped and torched with a high pressure gas torch. The mold can then be cooled, assembled, and fired prior to pouring.

The most well-known of these process variations is a development from the United Kingdom — the Shaw process. The chief difference between the Shaw and other investment molding processes is that a jelling agent is added to the refractory slurry-like mixture before it is poured over the pattern. When this mixture forms a somewhat flexible gel, the mold can be stripped off the pattern. Patterns can be made of various materials such as plaster, wood, or metal and may be reused. In this manner, this process differs from the expendable (wax or plastic) process. Molds are torched, then brought to a red heat in a furnace. The molds are allowed to cool prior to assembly for pouring.

Occasionally the Shaw process and the lost wax process are combined to gain the advantages of each. The complex pattern configurations which are difficult or impossible to remove from the mold can be made of wax and placed into the regular pattern. This provides for the regular pattern to be stripped off and the wax to be melted and burned out later.

When compared to investment castings the following apply:

Ceramic Mold Process Compared with Investment Casting

Casting Requirements	Ceramic Mold Castings	Investment Castings
Surface smoothness	80-125 microinch	40-125 microinch
Intricacy	Excellent, approaching but not equalling precision castings	Excellent
Thinness of metal sections	Excellent	Excellent
Tolerances	Good to excellent	Excellent
Machining costs	Machining greatly reduced, sometimes but not always eliminated	Minimum machining required
Lead time	Very short lead time	Longest lead time
Adaptability to various sizes	Casting size not restricted except above 100 pounds for the top size of the casting	Restricted to small castings
Adaptability to various metals and alloys	No limitations	No limitations
Pattern costs	Very low cost; job-bin wood or metal patterns may be used	Very high cost
Prototype adaptability	Low cost	High cost

1.4.7 Hitchiner Process (CLA, CLAS, CLAV)

The Hitchiner casting process utilizes a counter gravity (vacuum) system to fill the mold cavity with molten metal. The molds are usually produced using the resin-bonded shell and/or chemically bonded sand molding processes. The design of the mold provides small diameter feeders through the drag half of the mold to pull the metal up into the mold cavity. The mold is partially submerged into the metal bath and the metal is drawn into the mold cavities. The vacuum system draws the decomposition gases out of the mold cavity as they are generated during the pouring of the mold.

The process capabilities include the production of light sections castings produced in a variety of alloys normally not castable by other processes; castings with dimensional accuracy and casting finish normally only obtainable in the lost wax/investment process; casting integrity in alloy steels and nickel alloys in designs not normally castable by other processes, and castings up to 100 pounds in steel and high alloys including iron, nickel, and cobalt-base metals.

- *Advantages*

High casting yield, good casting definition, less cleaning of casting required compared to green sand, low sand and gas inclusions, and low scrap.

- *Disadvantages*

Low- to medium-volume production capabilities.

1.5 CHEMICALLY BONDED SAND MOLDING PROCESSES

This section describes the molding processes which use chemical binders mixed with sand to produce a mold. Since their introduction to the foundry, they have gained wide acceptance for the production of molds. Some chemical binder systems will bond or cure the mold through the chemical reaction established during the sand and binder mixing cycle, while other chemically bonded molding processes may require gassing of the sand mixture or heat applied to the sand mixture to complete the cure. The simplicity of the chemical system usually makes quality control less complex and results more consistent.

1.5.1 Shell Molding (Organic)

Probably the earliest most automated and most rapid of mold (and coremaking) processes was the heat-curing technique known as the shell process.

Resin-bonded silica sand is placed onto a heated pattern, for a predetermined time. Ejector pins enable the mold to be released from the pattern and the entire cycle is completed in seconds depending upon the shell thickness desired. The two halves of the mold, suitably cored, are glued and clamped together prior to the pouring of the metal. Shell molds may be stored for long periods if desired. Because of pattern costs this method is best suited to volume production. Designers should seek the advice of the foundry to ensure that all the benefits of the process are achieved.

- *Advantages*

Molds may be stored for extended periods of time. Good casting detail and dimensional accuracy. Mold is light in weight.

- *Disadvantages*

High pattern costs, high pattern wear (sometimes), high energy costs, higher material costs compared to green sand molding.

1.5.2 Sodium Silicate CO_2-Bonded Molding (Inorganic)

Instead of using an oil or resin that requires heat for bonding or curing, this process uses a sand which has been mixed with sodium silicate (NA_2SiO_3). In this process, the CO_2 gas forms a weak acid which hydrolyzes the sodium silicate, thus forming an amorphous silica which becomes the bond. There is also a bonding action from the sodium silicate itself. The use of CO_2 gives an almost instantaneous set. The mold is fully hardened before the pattern is drawn from the mold sections.

- *Advantages*

Low cost of materials. All sands can be used as the base aggregate for the silicate sand mixture. This includes silica sands, bank sands, lake sands as well as zircon sand, chromite sand, and olivine sand.

- *Disadvantages*

The more alkaline the binder, the longer it takes to gas and the greater the tendency for the core to remain rubbery instead of firm.

1.5.3 No-Bake Molding (Chemically Bonded Self-Setting Sand Mixtures) (Organic)

Since their introduction to the foundry industry, furan and urethane no-bake binders have gained wide acceptance for the production of molds and cores for casting. The advantages, regardless of which resins are used, are based upon control of the setting times by means of the addition of specific amounts of catalyst. This feature, combined with high strength and a desirable range of hot properties for use with various alloys presents the opportunity for great flexibility in mold making.

To produce a mold using no-bake binders, the pattern is covered to a depth of four to five inches with the no-bake sand mixture. The sand is allowed to set completely and the flask is then filled with conventional backup sand or, as the sand becomes tacky during its setting-up period, the backup sand may be added with a sand slinger. If ramming energy is applied before the sand sets, the density of the sand mass at the pattern surface is increased. This is reflected in better casting finish.

- *Advantages*

Simplified sand mix, good flowability of the sand mix. Uniform hardness throughout the mold. Exact dimensional control.

- *Disadvantages*

A limited bench life to the sand mixture. High sand temperatures are more critical. Patterns must be maintained in good condition.

1.6 SPECIAL AND INNOVATIVE MOLDING AND CASTING PROCESSES

This section describes several molding processes which tend to use special or innovative methods to produce a casting. Innovative metal casting processes are constantly being developed. Many of these processes are now proven practices and are being utilized for casting production involving many or all of the casting alloys.

Some of the molding and casting processes described are proprietary and may require licensing agreements. While as a technical society AFS avoids commercialism, it is felt that due to the success of the process or its originality of concept, a brief description will serve the primary purpose of this publication. Further details concerning any of the processes to be discussed may be obtained by contacting the AFS library.

1.6.1 Evaporative Pattern Casting

(Also referred to as Lost Foam Process, Expanded Polystyrene Molding, Full Mold Process)

The evaporative pattern casting process is an economical method for producing complex, close-tolerance castings using an expandable polystyrene pattern and unbonded sand. Expandable polystyrene is a thermoplastic material that can be molded into a variety of complex rigid shapes. Common uses for expandable polystyrene include insulation board, hot drink cups, novelty items, and packaging materials, as well as patterns for the foundry industry. The expertise for molding complex, thin-walled patterns with relatively smooth surfaces and at the low densities required for the casting process has also been achieved. With these developments, the evaporative casting process has made it feasible to produce high-volume production castings.

The evaporative casting process involves attaching expandable polystyrene patterns to an expandable polystyrene gating system and then applying a refractory coating to the total assembly. After the coating has been dried, the foam pattern assembly is positioned on several inches of loose dry sand in a vented flask and additional sand is then added while continuing to vibrate the flask until the pattern assembly is completely embedded in sand. A suitable downsprue is located above the gating system and sand is again added until it is level to the top of the sprue. Molten metal is poured into the sprue, vaporizing the foam polystyrene, and perfectly reproducing the pattern. Gases formed from the vaporized pattern permeate through the coating on the pattern, the sand, and finally through the flask vents.

In this evaporative casting process, a pattern refers to the expandable polystyrene or foamed polystyrene part that is vaporized by the molten metal. A pattern is required for each casting.

- *Advantages*
  Advantages of the evaporative casting process, as compared to the conventional bonded sand casting techniques, are numerous. Undoubtedly, the most important advantage is that no cores are required. Other advantages include: reduction in capital investment, reduces operating costs, castings can be made to closer tolerances as walls as thin as 0.120 inches have been cast, no binders or other additives are required for the sand which is reusable, flasks for containing the mold assembly are inexpensive, shakeout of the castings in unbonded sand is simplified — not requiring the heavy shakeout machinery, the need for skilled labor is greatly reduced, and casting cleaning is minimized since there are no parting lines or core fins.

- *Disadvantages*
  Pattern coating process is time-consuming, and pattern handling requires great care. Good process control is required as a scrapped casting means replacement not only of the mold but the pattern as well.

1.6.2 Vacuum ("V") Process Molding

This new adaptation of vacuum forming, the "V-process," permits molds to be made for the first time out of free-flowing, dry, unbonded sand without using high-pressure squeezing, jolting, slinging, or blowing as a means of compaction, which is generally used in conventional molding. The V-process is dimensionally consistent, economical, environmentally and ecologically acceptable, energy thrifty, versatile, and clean.

The molding medium is clean, dry, unbonded silica sand, which is "consolidated" by the application of a vacuum or negative pressure to the body of the sand. The patterns must be mounted on plates or boards and each board is perforated with vent holes connected to a vacuum chamber behind the board. A preheated sheet of highly flexible plastic material is draped over the pattern and board. When the vacuum is applied, the sheet clings closely to the pattern contours. Each part of the molding box is furnished with its own vacuum chamber connected to a series of hollow perforated flask bars. The pattern is stripped from the mold and the two halves assembled and cast with the vacuum on.

The original inventors of this proprietary process have established working agreements on a worldwide basis so that today individually licensed foundries using the V-process are producing castings of all sizes and shapes ranging from thin-sectioned curtain walls in aluminum to cast iron pressure pipe fittings as well as stainless steel valve bodies in weights up to massive 8-ton ship anchors. Other components being routinely cast include bathtubs, railroad bolsters and side frames, machine tools, engine parts, and agricultural castings. Any metal — gray, ductile, malleable iron, various grades of steel, or aluminum- and copper-base alloys — may be poured in a V-process mold with the possible exception of magnesium. The process is suited for castings varying from a few ounces to approximately 10 tons and permits the use of cores for complex configurations. The process is not intended for high productivity output, but its best application is in jobbing shops using flasks 30×30 inch.

- *Advantages*

  Superb finish. Good dimensional accuracy. Freedom from gas hole defects. Under ideal conditions can be highly competitive.

- *Disadvantages*

  Requires plated pattern equipment. Close synchronization of mold and metal readiness is essential in the foundry.

1.6.3 Centrifugal Process Molding

The essential feature of centrifugal casting is the introduction of molten metal into a mold which is rotated during solidification of the casting. The centrifugal force is relied upon for shaping and feeding the molten metal with the utmost of detail as the liquid metal is thrown by the force of gravity into the designed crevices and detail of the mold.

This method is ideally suited to the casting of cylindrical shapes but the outer shape may be modified by the use of special techniques. Normally, metal molds are used and molten metal is poured into the rotating mold which may spin about a horizontal, inclined, or vertical axis. The centrifugal force improves both homogeneity and accuracy.

- *Advantages*

  Improved homogeneity and accuracy in special circumstances.

- *Disadvantages*

  Limitations on shape of castings. Normally restricted to the production of cylindrical geometry shapes.

1.6.4 "H" Process Molding

This process utilizes a horizontal controlled pouring process for the repetitious production of castings. The "H" process utilizes rigid, double-sided molds produced by core blowing machines. Each mold carries a half casting impression on each side, complete with runner bar, feeder head, and ingate. Molds, when clamped together horizontally, form a pouring system of a top continuous runner and feeder bar. This runner/feeder bar is opened up at one end to form a pouring basin. Each mold is so designed that when the string of molds is poured, control of the flow of metal is obtained. The first mold is poured and is full before the metal proceeds to fill the second mold. This procedure of mold filling ensures a "flow feed" of each casting as all metal must pass through the feeder heads of the preceding castings. Maximum yield of metal is obtained. The horizontal clamping arrangement of the molds in the "H" process achieves the lowest possible ferrostatic casting pressure. This reduces the strain on mold parting joints, resulting in close control of casting dimensions and a minimum of casting cleaning. Presently the process has been applied to most cast metals.

- *Advantages*

  Close control of casting dimensions; reduction in casting cleaning costs; molds can be produced on core machines; molds can be produced from any of the known core or mold binder technology; high casting yield.

- *Disadvantages*

  Mold size limited to available equipment, the productivity controlled by existing production of molds, molding handling systems, and the pouring of mold assemblies.

1.7 A COMPARISON OF VARIOUS MOLDING AND CASTING PROCESSES

	Green Sand	Precision Molding			Chemically Bonded Molding
	Sand Casting	Permanent Mold Cast	Die Casting	Ceramic & Investment Casting	Shell, CO_2 No-Bake
Typical dimensional tolerances, inches	±.010"-±.030"	±.010"-±.050"	±.001"-±.015"	±.010"-±.020"	±.005" ±.015"
Relative cost in quantity	Low	Low	Lowest	Highest	Medium high
Relative cost for small number	Lowest	High	Highest	Medium	Medium high
Permissible weight of casting	Unlimited	100 lbs.	75 lbs.	Ozs. to 100 lbs.	Shell ozs.-250 lbs. / Co_2 & no-bake 1/2 lbs.-tons
Thinnest section castable, inches	1/10"	1/8"	1/32"	1/16"	1/10"
Relative surface finish	Fair to good	Good	Best	Very good	Shell-good CO_2-fair
Relative ease of casting complex design	Fair to good	Fair	Good	Best	Good
Relative ease of changing design in production	Best	Poor	Poorest	Fair	Fair
Range of alloys that can be cast	Unlimited	Aluminum-base and copper-base preferable	Aluminum-base preferable	Unlimited	Unlimited

(Characteristics are approximate and depend upon the metal.)

1.8 PARTING LINE INFLUENCE

When parting lines are considered, very close tolerances are difficult to obtain. A parting line absorbs fractions of inches per inch. A foundry is doing well to hold a parting line to 0.015 inch. Additional measurement is added to the casting tolerance.

2
Core Processes

One of the reasons castings are so widely used in all types of applications is due to the great design flexibility to form the internal shape of a casting which is inherent in a variety of processes available to produce a core.

The core, which is usually a pre-formed sand aggregate (metal for die casting), is inserted into the mold to shape the interior of a casting or that part of a casting which cannot be shaped by the pattern. The core is located and held in the mold by core pockets, formed by the pattern. In areas where the core may be thin in cross-section or require additional support to maintain metal wall thickness, chaplets are used.

In the core processes described below it is not possible to describe every known core production method, but the presentation of the processes listed is directed toward the major core binder systems which are presently used in casting production.

2.1 GREEN SAND CORE PROCESS

One of the oldest methods of producing a core is by using the same "green sand" used to produce the mold. The core may be produced by hand ramming, jolting, squeezing, or by blowing the core in a core blowing machine. Cores produced using green sand are simple in design and must be handled with great care. An advantage of green sand cores is that they do not require further processing (e.g., baking) once removed from the core box.

- *Advantages*

Cores and molds produced from the same sand system. No baking or curing of the core required.

- *Disadvantages*

Difficult to produce complex cores. Core must be handled carefully and requires constant care in handling.

2.2 OIL SAND CORE PROCESS

One of the oldest materials used to bond grains of core sand together is core oil. The composition of oil sand cores is generally 1% cereal binder and 1% core oil (by weight of sand) mixed with a clean (washed and dried) silica sand. The sand is mixed and blown on a core machine into a core box, or by hand ramming the sand mixture in the core box.

The core is then stripped onto a metal form (driers) or plate, which supports the core form during the baking cycle. The core should be baked for approximately one hour per inch of cross-section at a temperature from 375F (190C) to 450F (232C), depending upon the core oil manufacturer's recommendations. At the completion of the baking cycle, the core is strong enough to be handled for further processing. Its composition depends largely on the size and type of the castings which are to be produced, the production baking cycle, and the core oven facilities.

The oil sand core process is rapidly being replaced in the foundry industry by the chemically bonded coremaking processes which produce a core with better dimensional accuracy, improved physical properties, and higher productivity; however, there are many foundries still using oil sand cores in their casting process.

- *Advantages*

  Low material cost. Improved core dimensional accuracy over green sand cores. Cores can be handled and stored for a limited time allowing flexibility in production scheduling which is not possible when using green sand cores.

- *Disadvantages*

  Dimensional accuracy is not consistent. Low productivity.

2.3 CHEMICALLY BONDED CORE PROCESSES

This section provides a brief description of a variety of chemically bonded coremaking processes.

The two major methods used to transfer sand to a core box are by blowing the sand mixture into a closed core box, or by manually, mechanically, or pneumatically feeding the sand mixture into an open core box.

The methods and materials used to fill the core box cavity and produce the chemical reaction necessary to complete the curing of the core binder system vary significantly.

In the first section high-volume production systems (blowing into a closed core box) are described, followed by medium- to low-volume production systems (feeding the sand into an open core box), and a listing of the various chemical binder systems.

2.3.1 High-Volume Production Systems

The majority of high-volume production systems produce a core by blowing the sand into a closed core box. The sand has been mixed prior to transferring to the core blowing machine. The completion of the curing cycle depends upon the core binder requirement for heat, gas, or a chemical reaction of the resin and catalyst to complete the cure (hardening) of the core before the core box can be removed.

A gas-purge system requires more equipment to complete the gas and purge cycle to producing a core on the cold box system than a heating system required to produce a core on the hot box system. The application of heating or gas-purge systems vary in cost when using a three-part binder system.

Some of the three-part binder systems can produce a core without the application of heat or gas. This machine has three small sand mixers located on the top of the machine which feed the mixed sand into the sand magazine for flowing into the core box. After blending of the total resins and catalyst with sand a very fast cure occurs. This system requires very close process control during the blending and mixing cycle, but does not require the added equipment required in the other two methods described above.

The materials used for core boxes for the above production systems are usually metal or plastics due to the high-volume requirements. Lower volume requirements make it possible to use wood or plastic patterns for the gas-purge system or the three-part fast cure systems.

2.3.2 Medium- to Low-Volume Production Systems

The equipment required to service a medium- to low-volume core production system has less equipment and usually the equipment is not as complex as a high-volume production system. The core boxes are usually designed as a "split" core box which is open on the top to receive the sand from a screw type tube mixer. The sand flows out of the mixer into the core box which is usually mounted on a compaction table required to produce a dense core. After filling the core box, the overflow sand is "struck-off" to the top of the core box. The core box is then moved to an area provided for curing. Curing time is dependent upon the binder system and the cross-section of the core.

This process is used extensively to produce large cores. One of the disadvantages of this process is the pasting of cores together since most cores must be produced in sections.

2.3.3 Listing of Various Chemical Binder Systems

Chemical binder systems can be further categorized as organic or inorganic, and thermosetting, self-setting, or gas-cured. A list of the major core binder systems which are presently used in the foundry industry in the production of castings will be found on the next page.

Chemical Binders

- ORGANIC
  - Thermosetting processes
    - Shell:
    - Hot box (furan/phenolic):
    - Warm box:
    - Core oil:
  - Self-setting systems
    - Furan:
      - High-nitrogen furan-acid
      - Medium-nitrogen furan-acid
      - Low-nitrogen furan-acid
    - Phenolic:
      - Phenolic-ester cured
      - Phenolic-acid
    - Urethanes:
      - Alkyd-organometallic (alkyd-oil)
      - Phenolic-amine
      - Polyol-amine
  - Gas-cured processes
    - Free radical
    - Phenolic urethane-amine
    - Furan/peroxides-SO_2
    - Epoxy/SO_2
    - Phenolic-ester cured

- INORGANIC
  - Self-setting processes
    - Silicates
    - Sodium silicate-ester cured
  - Gas-cured processes
    - Sodium silicate-CO_2

For further information regarding various binder systems contact the American Foundrymen's Society.

The advantages and disadvantages of chemically bonded systems when compared with oil sand cures are noted as follows:

- *Advantages*

Capital expenditures are reduced; higher cured strength at room temperature; reduced skill required; less damage in handling and storage; core storage for longer periods of time; better dimensional accuracy; higher productivity; chemically bonded systems usually makes quality control less complex and results more consistent.

■ *Disadvantages*
Costs of the various binders are higher; process control must be consistent.

3
Melting Processes

3.1 THE CUPOLA MELTING PROCESS

The cupola is the most economical melting unit for cast iron. With proper metallurgical control, it is quite adequate for melting all the grades of gray irons.

The cupola is a simple stack having a steel shell closed at the bottom and opened at the top; a lining proper of fire brick; a rammed-in or daubed-on lining containing fireclay, a door for charging fuel and metal, opening for air-blast, spouts for drawing off metal and slag and bottom doors.

Successful cupola operation hinges largely on combustion control. In order to have favorable melting conditions, it is necessary to have a balanced combination of coke and air supplied at the proper rate. With proper combustion, the control of metal composition, temperature, and slagging can be accomplished.

Due to the fact that the air necessary to consume the fuel and melt down the metal passes up through many feet of broken up metal and coke, natural draft will not suffice. Air must be supplied under some pressure and introduced as near to the bottom as the desired level of molten iron will permit. It is necessary that the air enter at several openings, through the cast iron ducts called *tuyeres*, in order to secure uniform distribution through the fuel. The tuyere is usually flaring in shape, the large end being on the inside in order to make the blast as nearly uniform as possible, and because the coke of the fuel bed interferes to a great extent with air entering through the openings.

The cupola is divided into five zones: the crucible or hearth, the tuyere, melting zone, charging zone, and the stack. The hearth or well extends from the sand bottom to the bottom of the tuyeres. The tuyere zone is the section where the blast enters and it extends from the top of the hearth to the melting zone. The melting zone is where the melting takes place and is the hottest part of the cupola. The charging zone extends from the melting zone to the charging doors. The material in this zone absorbs heat, thus aiding the melting process. The stack zone extends from the charging doors to the top of the cupola.

3.2 ELECTRIC FURNACE MELTING

Direct Arc Furnaces

Direct arc furnaces are used mainly for melting ferrous metals (steels), although some nonferrous metals are also melted.

In the direct arc furnace, the body of the furnace is enclosed in a refractory-lined cylindrical steel shell, pivoted or mounted on rockers so that it may be tipped forward and backward. It is provided with three vertical electrodes, each connected to one phase of a three-phase alternating current circuit. The furnace roof is built into a separate steel ring and rests upon the side walls of the furance shell.

The electrodes are suspended from horizontal arms which are attached to structural steel masts and extend through openings in the furnace roof, provided with water-cooled rings.

The furnace usually has a working door on each side and a taphole and spout at the front. The charge is placed into the furnace through the roof which can quickly be swung off the body of the furnace and to one side in order to allow the charge to be lowered into the furnace. The entire charge, contained in a drop-bottom bucket, is then lowered into position with the overhead crane.

Induction Furnaces

There are basically two types of induction furnaces available: the core type or channel furnace, and the coreless or crucible furnace.

The *channel induction furance* is the oldest induction furnace. It consists of an upper case in which the metal is melted and stored. The power section, or inductor is attached to the bottom of the furnace. It contains a transformer coil and core with the metal loop or channel forming the secondary of that transformer. It is here that the heat necessary to do the melting is created.

Stirring action carries the heat to the main bath by moving the metal upward and conduction takes care of the rest. This furnace is never turned off, and never emptied except to reline either the inductor or the upper case. It must be started with molten metal. Alloy changes can be made, but only within compatible ranges. It is used for either melting or holding ferrous or nonferrous metals.

The *coreless induction furnace* can be started with a cold charge, poured empty, and can be of a smaller capacity than the channel furnace.

Induction furnaces are further distinguished by the frequencies at which they operate.

The *high frequency induction furnaces* operate on the principle of an air-cored transformer with the charge representing the short-circuited secondary winding of a single turn coil. High frequency power is used to decrease current penetration and increase heating effect. High frequency induction furnaces are especially suited for the melting of odd lot and specialty bronzes as well as for quality bronzes.

In the *line frequency coreless furnace*, a slug or heel is used to effect coupling when the melting cycle begins. Although the furnace can be completely emptied, it is wise to leave a heel to speed future melts when the same alloy is to be run.

The *high frequency coreless furnace* can be charged with ingot stock scrap. This furnace is ideally suited for small heats and where fast melting and alloy changes are required. As the charge melts down, the remainder of the addition can be added. As with all induction furnaces, melting losses are low, and charge additions are quickly taken into the bath because of the inherent electromagnetic stirring action which provides for uniform composition throughout the bath.

When starting a *low frequency channel furnace* a heel of molten metal equal to a fourth of the charge must be poured into the channel to complete the secondary circuit. Borings, sprue, ingots, or returns can be added when the charge is melted. Other additions such as tin, lead, and zinc may be added just before pouring. When proper pouring temperature is attained, top slag is skimmed off. The furnace is never completely emptied, as a molten heel must be kept in the crucible and above the channel to maintain melting production.

Electric Resistance Furnaces

Electric resistance furnaces obtain the necessary heat by running an electric current through one or more resistor rods (electrodes) of carbon or silicon carbide. The rods do not touch the metal. Current passes through the electrodes which strike an arc in the center of the melting chamber. The molten metal is not part of the electrical circuit as is the case in the induction furnace. Electric resistance furnaces are used as melting and holding furnaces in nonferrous operations.

3.3 CRUCIBLE MELTING FURNACES

Crucible melting is the oldest and simplest way of melting metal. The crucible furnace may be fired with coke, oil, gas, or electricity which heats the metal through the walls of metal or refractory crucibles made of clay-graphite or silicon carbide bonded with carbon.

Crucible furnaces are of three types: liftout, stationary, and tilting. In the *liftout (pit) furnace*, the crucible is placed in a furnace shell. After melting, the crucible is removed and taken to the mold for pouring. Capacities for this type of crucible range up to about 100 pounds. The operation is flexible with respect to change of alloys, ease of treating the melt, and the simplicity in which metals can be provided at any desired temperature. Capacity is limited. Equipment cost is probably the lowest.

Stationary crucible furnaces have fixed crucibles which are rarely removed until replacement is necessary. Capacities range from 300 to 1,000 pounds. The crucibles are refractory or cast iron.

Metal is removed from these furnaces by dipping more often with hand ladles. Such furnaces are usually batch melters or holding furnaces. When used as a holding unit, the batch is kept at constant temperature and supplied with molten metal from another source.

Cast iron crucibles are mostly used for holding molten aluminum at constant temperatures with relatively low heat input.

Both cast iron and refractory crucibles are used in *tilting furnaces* with capacities of about 1,000 pounds. The furnace mounted on trunnions can be tilted so that the molten metal pours into the transfer ladles. There are two types of tilting furnaces. One pivots on a horizontal axis at its midpoint so that the lip moves up and down. The other, known as a nose-pouring or nose-tilting, pivots about the pouring spout, which remains stationary.

Tilting is manual, electric, or hydraulic. This type of furnace has large capacity, fast melting characteristics, and its design simplifies metal transfer to the pouring ladles. The disadvantage of all tilting furnaces is that the metal is severely and harmfully agitated during transfer. This agitation mixes air with the metal and leads to inclusions of both aluminum oxide and gas.

3.4 REVERBERATORY MELTING FURNACES

The reverberatory furnace is constructed of refractory brick in a steel structure. The metal, lying in a shallow open bath, is heated by flames and hot gases from the combustion of gas or oil. Occasionally, heat is supplied by *electrical resistance* radiation units in the furnace roof, though this method is almost entirely confined to holding furnaces.

There are many types of reverberatory furnaces distinguished by methods of charging and removal of the metal. In one type, metal is charged to the main melt, in another, into a puddler or charging well exposed to the air at one side of the furnace. In another type, it is charged through the stack, taking some preheat from gases of combustion leaving the furnace. In still another type, it is charged on a sloping dry hearth and then, after melting, runs into the reverberatory holding chamber. This type of furnace — also known as a two-chamber dry-hearth furnace — has some outstanding advantages; for example, all meltdown dross remains on the dry hearth and makes no contact with the molten metal in the second chamber. Combustion control maintains an oxide skin in the molten metal and cuts gas absorption. Freedom from agitation in the holding chamber gives maximum assurance of clean metal. Metal is removed in three ways: through a pouring spout by tilting the furnace, by ladling or dipping from wells at one side of the furnace, or by pouring through transfer ladles through a tap hole below the surface of the melt.

Reverberatory furnaces range in capacity from several hundred to 120,000 pounds and higher. These furnaces are generally used for producing large quantities of a single alloy at a uniform pouring temperature. They produce at minimum cost and are adapted to a broad variety of operations. As to disadvantages — they are not flexible in making changes of alloys or pouring temperatures and are difficult to work in alloying, fluxing, and degassing.

4

Types of Metals and Alloys

This section describes the metals used to produce castings in the various casting and molding processes. The selection of a metal is determined by several factors which include the characteristics of strength, ductility, dimensional accuracy, machineability, resistance to corrosion, as well as castability. This section describes the basic metals most commonly used in the industry today, but does not attempt to describe the metals which, by altering the chemical composition, change the base metal into a special metal with specific characteristics. Cast metals are generally classified into two groups: ferrous and nonferrous.

4.1 FERROUS METALS

Ferrous metals may be subdivided according to carbon content and classified as steel or cast iron. Typical casting groups classified as ferrous grades of metal are as follows:

4.1.1 Gray Iron

Gray iron castings can be cast in various tensile strength ranges from soft, machineable, low strength irons of near 20,000 psi tensile strength to hard, wear-resistant irons of 60,000 psi. Various casting markets are served by individual classifications of gray iron, noted as follows:

Class 20	*Class 30*	*Class 40*	*Class 50*
Ingot mold	Auto (35)	Valves	Engine
Municipal	Farm	Mach. tool	Spec. ind.
Soil pipe	Constr.	Gears	Pumps
Motors	House app.	Machinery	Compr.

Class 60	*White Iron Ni-Hard, High Cr.*	*Alloyed Irons, Ni-Resist*	*Compacted Graphite Irons*
Engine	Metalwkg.	Engine	Molds and stools
Mach. tool	Cement	Pmps	Exhaust manifolds
Mining	Rolls	Food prod.	Flywheels
Gears	Coal pulver.	Marine	Axle housings
			Hydraulic valves

4.1.2 Ductile Iron

Ductile iron castings can be produced in various grades and types, ranging from 60,000 psi in tensile strength to over 150,000 psi, and from 1% to 25% elongation. Various market segments and casting types are served by the following grades:

Ferritic	*Pearlitic/ Ferritic*	*Pearlitic*	*Martensitic*	*Bainitic*
60-40-15 60-45-12 60-40-18	80-55-06 80-60-03	100-70-03	120-90-02	130-100-04 (Average)
Valves Pipe Farm Constr. Pumps Elec. Motor veh.	Engine parts Farm Constr. Mach. tool Spec. ind. Refrig. Motor veh.	Farm Mach. tool Gears Refrig. Motor veh.	Rolls Dies, tools Gears	Gears Shafts

4.1.3 Malleable Iron

Malleable iron is a desirable engineering material because of its ease of machineability, its toughness, ductility, and wide range of strengths. Some of the principal industries using castings made of malleable iron are: automotive, hand tools, and valves and fittings. Malleable iron castings continue to decline in usage due to the competitive advantages of ductile iron.

4.2 STEEL

Steel is an alloy of iron and carbon that may contain other elements and in which the carbon content does not exceed approximately 1.7%.

Steel is considered the ideal metal for many types of casting applications because its chemical analysis as well as its mechanical and physical properties are easily varied over a wider range than other cast metals. This is achieved by varying the carbon content, the chemical composition, or by heat treatment. It is strong, tough, dependable, and readily joined by welding or bolting to other metal forms, such as rolled products, forgings, or other castings. Steel castings are used in many industries. These include railroad, automotive, marine, farm equipment, machinery for highway, construction, mining, metalworking, power transmission, valves and fittings.

4.2.1 Carbon and Low Alloy Steel Castings

Markets for carbon and low alloy steel castings include the following:

Valves
Turbine
Construction
Mining
Oil field
Mill and metalworking machinery
Pumps
Trucks

4.2.2 Corrosion-Resistant Steel Castings

Corrosion-resistant steel castings include all stainless steel castings used in the following markets:

Ship propellers
Ship and boat building
Special industrial machine components
Valves and fittings
Internal combustion engine turbines
Oil field machinery and equipment
Roll mill machinery

4.2.3 Heat-Resistant Steel Castings

Heat-resistant steel castings include centrifugally and statically cast alloys, heat-resistant iron chromium alloys, and iron-chromium-nickel alloys.

A summary of the heat-resistant steel markets is as follows:

Nonelectric heating
Industrial furnaces and ovens
Boiler shop production
Blowers and fans
Valves
Tools, dies
Mining
Rolling mill machines
Oil field equipment

4.2.4 Manganese Wear-Resistant Steel Castings

Manganese alloy steels, which cannot be made except by the casting process, have excellent resistance to wear; they work-harden during use.

Potential markets include:

Construction and mining equipment
Special industrial machinery
Railroad equipment

4.3 NONFERROUS METALS

Nonferrous alloys are classified according to the base elements of which they are composed. The base elements used commercially are mainly copper, aluminum, magnesium, lead, tin, and zinc. To obtain the desired physical and mechanical properties, it is necessary to vary the amounts of these elements. Typical nonferrous grades of metal most commonly used are as follows:

4.3.1 Brass

Copper-base alloys of copper and zinc are commonly classified as *brass*. Copper-zinc alloys are the major group of metals used, due to their desirable properties and relatively low cost.

Yellow brass is the most ductile of all the brasses. Its ductility makes possible the use of this alloy for jobs requiring the most severe cold-forging operations, such as deep-drawing, stamping, and spinning.

Red brass is composed of 2% to 8% zinc, has a reddish color, a great resistance to corrosion, and good workability. Red brass alloys have good casting and machining characteristics. They are readily shaped by stamping, drawing, forging, and spinning. Applications of red brass include valves, fittings, rivets, radiator cores, plumbing pipe, flexible hose, and screen cloth.

4.3.2 Bronze

When tin is added to copper as a secondary alloying element, *bronze* is produced. Hardness and wear resistance are increased, although ductility is decreased. Tin bronzes have excellent resistance to corrosion and good properties at elevated temperatures. When lead is added an excellent *bearing bronze* is produced which is useful as a bearing material.

Yellow bronze, with more than 17% zinc and more than 2% aluminum, manganese, iron, and tin, is known as *manganese bronze*. Manganese and iron, which are added to obtain the desired physical properties, can increase tensile strengths from 60,000 to 120,000 psi.

Copper-rich alloys of aluminum and copper are known as *aluminum bronze*, which has a near-gold color, possesses a good finish, and is highly resistant to acid.

One of the many grades of bronze, which is red in color and contains 1% to 5% silicon, is known as *silicon bronze*. This alloy has good resistance to salt water, alkalies, and acids. Small amounts of zinc, tin, and iron may also be present.

4.3.3 Nickel-Base Alloys

Nickel-base alloys have great resistance to corrosion in the presence of most mineral acids, most organic acids, and all alkalies. They are not resistant to corrosion by nitric acid or by oxidizing salts. They have good mechanical strength, ductility, and resistance to wear, although they cannot be used for bearings, except under light loads and at slow speeds.

An alloy of 70% nickel and 30% copper is known as *Monel*. When silicon is added to it, it becomes age-hardened, and thus more wear-resistant. An alloy of 80% nickel and 20% chromium, known as *Nichrome*, is used for electrical resistance heaters. An alloy of 80% nickel, 14% chromium, and 6% iron, known as *Inconel*, is used where oxidation resistance with high strength at elevated temperatures is needed.

Alloys with a high percentage of nickel are used for chemical equipment, such as implements used in the dyeing of textiles and the manufacture of caustics, as well as for making water-softening equipment, valves and pump parts, and food-handling equipment.

An alloy consisting of 20% zinc, 20% manganese, 1% aluminum, and the balance copper produces a *white-bronze* casting that is strong, ductile, and corrosion-resistant. It can be polished to a high, silvery luster which makes it useful in architectural and marine hardware, plumbing fixtures, ornamental castings, hospital equipment, and swimming pool equipment.

Casting markets using the brass, bronze, copper, and copper-base alloys include the following:

Plumbing valves and sanitary fixtures and fittings
Pumps
Electrical equipment and apparatus
Power transmission equipment
Marine hardware and ship propellers
Metalworking machinery and equipment
Ball and roller bearings

4.3.4 Zinc-Base Alloys

Zinc-base alloys are widely used in die casting. The alloying elements are principally copper, aluminum, and magnesium. The amount of each used depends on the properties desired. High purity zinc is used as the base metal. The addition of copper increases the strength, but reduces ductility. Addition of aluminum improves the strength of the alloy and delays the rate of attack of the alloy on steel dies, and thus improves the life of the die. Additions of magnesium add to the dimensional stability of a die casting.

Alloy development is a continuous process and the producers should be contacted when there is a question regarding the alloy to be used with the casting process being considered.

Zinc die casting markets include the following:

Bathroom and plumbing fixtures
Door, window and furniture hardware
Hand tools

4.3.5 Aluminum Alloys

Pure aluminum is a poor casting material and is limited to the casting of rotors where high electrical conductivity is an advantage. Almost all aluminum castings are made of an alloy of aluminum. The selection of a particular alloy depends on the criteria demanded of the alloy — the mechanical strength, machineability, surface appearance, resistance to corrosion, conductivity, pressure tightness, and other factors. The principal alloying elements are copper, silicon, magnesium, zinc, chromium, manganese, tin, and titanium. Iron is often present in small quantities, and is usually considered an impurity. Although all of the alloys are commercially castable by a specified process, namely sand casting, permanent mold casting, or die casting, the casting industry has established a preference for certain alloys in each casting process.

4.3.6 Sand Casting and Permanent Mold Alloys

In sand casting and permanent mold, alloys with additions of silicon in the range of from 5% to 8%, copper in the 1% to 5% range, and magnesium in the 0.2% to 1% range are used to produce most aluminum castings.

Sand cast and permanent mold aluminum casting markets are noted as follows:

Internal combustion engine
Lawn and garden equipment
Power hand tools
Office machines and computers
Aircraft parts
Household appliances

4.3.7 Die-Casting Alloys

In die casting, approximately 80% of the castings produced use an alloy with an addition of 2% to 5% copper, 7% to 10% silicon, 1% to 3% zinc, and 1% iron.

Castings produced using the die-casting process are used for the following:

Automotive and truck combustion engines
Furniture, fixtures household appliances
Light equipment, communications
Toys, sporting goods, bicycles, motorcycles
Motors, generators, regulators, instruments

4.3.8 Aluminum-Magnesium Alloys

Aluminum-magnesium alloys are characterized by excellent mechanical properties and machineability, as well as resistance to corrosion. They have good resistance to impact and high ductility.

These alloys are used in:

Marine applications
Highway fixtures

4.3.9 Magnesium Alloys

Magnesium high purity alloys with various amounts of aluminum, zinc, zirconium, and rare earths are used to produce sand castings, permanent mold castings, and die casting, with die casting the predominant method.

Magnesium die casting markets include the following:

Aircraft components
Power tools and sporting goods
Automotive — transmission cases, clutch housings, timing chain covers, wheels, hinges, brackets, brake pedals, head lamp frames, oil filter adapters, etc.

4.1 PREMIUM QUALITY CASTINGS

Utilizing conventional foundry processes in a selective manner produces a premium quality casting, which applies a combination of features not found in any method used to produce these castings. Techniques have been developed to meet the exacting demands of the military, electronic, aerospace, and computer industries for quality and reliability.

4.4.1 Comparing Dimensional Tolerances

Typical dimensional tolerances for ferrous and nonferrous castings vary considerably from process to process. A comparison of dimensional tolerances will be found on the next page.

Typical Cast Metal Dimensional Tolerances

	Ferrous Metal		Nonferrous Metal	
Process	Steel	Gray Iron	Copper-Base Alloys	Light Alloys[2]
1″-6″	1″-6″	1″-6″	1″-6″	
Green Sand Molds[1]	.060-.100	.030-.050	.015-.015	.015-.015
Shell Molding[1]	.020-.050	.019-.040	.016-.019	.010-.014
Permanent Molds[1]	.015-.020	.030-.035	.018-.025	.009-.012
Die Casting[1]	—	—	.005-.080	.005-.080
Investment Casting[1]	.010-.020	.010-.020	.005-.010	.005-.010

[1]plus or minus inches
[2]aluminum/magnesium/zinc-base alloys
[3]these tolerances have no parting lines. If the casting passes a parting line, add 0.015 in. -0.020 in. for a one-inch section casting. If the dimension is between a core and mold surface, the difference is approximately the same.

5
Casting Design

5.1 DESIGN CONSIDERATIONS

With varying degrees of difficulty, almost any shape can be cast. If one ignores cost and quality, the range of foundry processes permits tremendous flexibility and complexity of design. This is both an advantage and a disadvantage to the casting user, because it is so easy to produce a casting design which is difficult to cast completely free of defects.

Successful castings buying begins with good design, coupled with a wide choice of foundry process. Companies using large quantities of castings should employ a competent castings engineer. Smaller companies should encourage their foundry suppliers to provide design suggestions to improve qualtiy, reliability, and reduce cost. Impartial advice may be obtained from consultancy services.

For best results, casting design advice should be applied in the *early* stages of the design. Cost reduction and/or quality and delivery improvement may result from minor changes in existing designs which do not affect the function of the casting.

Buyers should encourage foundries to suggest small changes which eliminate problems or facilitate production.

It is not within the scope of this guide to identify the problems of casting design. It is a broad subject dealt with in depth in several text books for those whose primary responsibility is in the casting design field.

The buyer may, however, benefit from the knowledge of a few simple rules.

1. Rounded surfaces and generous radii produce better castings.
2. Extensive flat surfaces are difficult to cast accurately. Strengthening ribs are desirable.
3. Rapid changes in section thickness should be avoided.
4. Material selection and the correct alloy composition used for any given metal thickness is an important design criteria.
5. The shape of the castings should avoid shrinkage cavities.

6. It is desirable to draw the pattern from the sand without the use of loose pieces. In some low volume jobs, however, loose pieces may be the cheapest way of achieving the desired shape.
7. Although cored holes and apertures often reduce machining costs, it is virtually impossible to avoid core flash where the cast and cored surfaces meet. Cores should be balanced and properly supported, preferably at both ends and, if possible, at the sides.
8. Wherever possible, bosses should be designed into the mold joint line.
9. Cores can be used to avoid changes of section. A solid casting is not always the preferred design.
10. Adding thickness or weight to a casting does not necessarily increase strength or give more reliable quality.
11. A solid casting is difficult to produce free of defects and to reliable quality standards.
12. The use of core setting fixtures involving the placing of cores into blind coreprints should be avoided.

5.2 DESIGN FEATURES OF VARIOUS TYPES OF CASTINGS

Sand Castings

There is a wide choice of ferrous and nonferrous materials, and considerable complexity is possible. The size is limited only by melting capacity. Section thickness depends on the material used but 1/8 in. should be regarded as the absolute minimum, with 5/16 in. the desirable minimum for most popular grades of cast iron. The tolerences are from plus/minus 1/16 in. to plus/minus 1/4 in. for iron castings over 30 in.

The figures for aluminum alloys and copper alloys are similar to gray iron but it is impossible to generalize because of the wide range of alloys available.

The minimum section thickness for steel is 1/4 in. Tolerances depend on physical shape, precise metallurgy, size of casting, and class of pattern equipment to be used.

Where doubt exists, it is recommended the buyer consider the advice of the foundry in deciding the practical limits for any given design.

Shell Molded Castings

There is a wide selection of ferrous and nonferrous materials. Complexity is restricted by factors such as cycle times, and ease of withdrawal of the mold from the pattern. The process is mainly suitable for small castings weighing only a few pounds. Larger, heavier castings can be produced by using sand or shot backing of the shell molded sections. Accuracy is good. Tolerance is plus/minus 0.010 in. per inch. Under some conditions, section thickness may be slightly less than for sandcastings.

Diecastings (Permanent Molds)
There is a limited choice of materials, normally restricted to zinc and certain light alloys. Iron and bronze can be cast in permanent molds if simple shapes are used. Low melting point alloys provide considerable design flexibility and permit great complexity. High production rates can be achieved with pressure diecastings.

Tolerances vary from plus/minus 0.0015 in. per inch for low melting point alloys to 0.004 in. per inch for cast iron or bronze. Minimum wall thickness of as little as 0.050 in.-0.070 in. is possible in some light alloys.

5.3 GENERAL RULES — ALL TYPES OF CASTINGS

1. Specify close tolerances only where necessary.
2. Avoid heavy and very thin sections wherever possible.
3. Recognize that changes of section promote stresses during cooling and will affect overall tolerances.

6
Pattern Equipment

6.1 ECONOMICS OF GOOD PATTERN EQUIPMENT

With the possible exception of those companies involved in high-volume production and accustomed to a high tooling investment, there is a general reluctance to spend money on pattern equipment.

Castings buyers frequently attempt to reduce the initial cost, choosing wooden patterns where metal is essential or selecting low-cost plastic patterns instead of the new urethane pattern materials. Buyers may opt for part cost patterns, which is not recommended as full ownership of patterns is desirable.

As a general rule, economies on patterns are unwise. The buyer should negotiate attractive prices for patterns, and avoid the temptation to downgrade the pattern equipment in the interests of lowering the initial cost. Quality pattern equipment constructed to a high standard, consistent with the demand for castings, the accuracy required, and well maintained, is necessary for quality castings at minimum cost. The precision and accuracy of premium quality castings depends largely on the care with which models and patterns are made. Mold materials and systems have been developed which can produce close dimensional accuracy and smooth surface finishes. Plaster or metal molds are often used to produce close dimensional tolerances and smooth surface finish. Expandable wax or plastic patterns are used to produce intricate parts.

Having patterns constructed by a pattern maker without consulting the foundry is another hazard to avoid. The foundry may add a small mark-up to any patterns they purchase, but usually a lower cost results by placing the full responsibility with the foundry when they provide the pattern equipment.

In pursuing this course, the buyer may feel at risk should the need arise to transfer equipment from one foundry to another. If the foundry choice is correct, this need will seldom arise. If it is necessary to provide this kind of flexibility, and the patterns are well made, it is a fairly inexpensive process to remount patterns onto new boards or pattern plates to suit the new foundry.

6.2 CLASSES OF PATTERN EQUIPMENT

There are many types of pattern equipment.

1. Inexpensive wood patterns for prototypes, possibly loose-molded.
2. Hardwood patterns mounted on boards for regular low-volume or small-batch quantity production.
3. Plastic pattern equipment for medium-volume production where reasonable wearing properties are important. Some plastic has the added advantage that it is shrink-free and therefore ideally suited for the reproduction of wooden patterns if their condition is suitable.
4. Metal patterns are produced in cast iron or cast aluminum according to the foundry or process used. Brass or bronze patterns may also be used.
5. Dies for lost wax castings are often single impression and machined from the solid. It is possible to use master patterns (replicas of the casting to be produced) to facilitate the manufacture of lost wax dies, but advice from the foundry should be sought in each individual case.
6. Dies for pressure and gravity die castings may be machined from the solid stock or may be cast to shape and then finished to the final dimensions. These are generally very complex and expensive pieces of equipment involving moving cores, etc., with a high degree of skill in both the design and manufacture required.
7. Shell cores are sometimes used where dimensional accuracy is important and shell coreboxes may be used for producing cores for castings made from wooden patterns, however, shell cores are normally used with sand- or shell-molded castings where metal patterns are in use. The method of producing shell cores relies on heated coreboxes and a sand/resin mixture being introduced into the cavity of the desired shape.

The advantages and disadvantages of the different classes of pattern equipment are summarized in the following table.

Inexpensive Wood Loose Patterns

- *Advantages*

Relatively low cost to produce. Can be used to produce plastic patterns. Easily modified.

- *Disadvantages*

Not suitable for production batches unless converted into plastic before becoming worn. Tend to become inaccurate and to produce poor surface finish.

Hardwood Patterns

- *Advantages*

Relatively inexpensive if required for small-batch production. Can be used as masters for plastic patterns. Fairly easily modified but expensive if multiple impressions involved.

- *Disadvantages*

Fair wearing properties. With care, will remain reasonably accurate over fairly long periods. Not suitable for high-volume production.

Plastic Patterns

- *Advantages*

Relatively inexpensive to produce. Urethane technology provides good wearing properties and accuracy. Easily modified depending upon the materials used. Suitable for high-volume production.

- *Disadvantages*

New urethane technology is now producing high-volume production patterns which produce castings with the same dimensional tolerances as metal patterns and has reduced many of the disadvantages of the early plastic pattern materials used.

Metal Patterns

- *Advantages*

Excellent wearing properties. High standard of accuracy and stability. Suitable for high-volume production and shell molding.

- *Disadvantages*

Costly. Difficult to modify. Equipment less readily adaptable for movement from one foundry to another. CAD/CAM improvements are greatly reducing cost and lead time requirements.

6.3 OWNERSHIP OF PATTERNS AND DIES

The main points to consider are:

1. Full ownership gives the buyer the right to withdraw equipment and transfer to another foundry.
2. Full ownership of dies gives the right of transfer, but there may be technical problems moving equipment to another foundry.
3. Normally die-cast foundries offer only part cost of tooling, thus retaining an interest in the equipment.

4. When patterns are amortized, the buyer should ensure that the pattern cost element in the unit cost is deducted when the agreed quantity has been produced.

5. In those cases where the foundry agrees to provide patterns at no cost to the buyer, the unit cost should be adjusted when initial pattern costs and interest charges have been paid off.

7 Purchasing Policy for Castings

7.1 SELECTING THE FOUNDRY

There are many ways to select foundry suppliers: names taken from a buyers guide, from the visitors' book, advertising material, magazines, and recommendations from other buyers and engineering staff.

Buyers should actively seek more reliable information about sources of supply and try to match foundry capacity to their purchasing needs.

This can be achieved by issuing a questionnaire to prospective suppliers. An example of a questionnaire is shown on page 48 through page 50. The answers provide the buyer with data concerning his potential suppliers. Foundries may be reluctant to answer some of these questions; however, the buyer should insist this is done as a prerequisite to consideration for future business.

Replies should be studied with care because they will reduce the scope for error and provide the guidelines for future purchasing policy. For example, the buyer will know whether the supplier can produce the grade of material required in sufficient quantity. Attitudes on quality can be assessed from the percentage of total production and quality control personnel directly involved in the quality control program. The buyer should recognize that a foundry's strong emphasis on quality control indicates they are probably producing very high grade castings. *Their price structure will reflect this whether or not castings to exceptionally high standards are required.*

If it is intended to buy heavily cored castings, look for the supplier who produces this type of casting. Discuss capacity to avoid disappointment after the order is placed. Failure to make delivery schedules may be the result of coremaking capacity.

Generally speaking, buyers would do well to make buying decisions on the realities of the situation: the price, quality/inspection standards required for any given set of physical and mechanical properties.

From the answers to the questionnaire, the buyer may ascertain whether or not transportation will be a problem and the prospects which may exist for further development of business with the supplier. Perhaps the most important point, when all other conditions have been satisfied, is to look at the foundry's current pattern of production. Is it high volume, small castings, or jobbing and what are their *standard flask sizes*? Flask size can have an effect on prices and quality of castings. If possible, when designing castings, keep in mind the flask size, so as to maximize the number of patterns on a plate. This will be a significant step in the process of cost reduction. Always allow sufficient space around the casting for the coreprints (the projections needed to provide adequate support for the core). Consult the foundry supplier if in doubt on this point.

7.2 THE FOUNDRY AS AN EXTENSION OF THE BUYER'S MANUFACTURING CAPABILITY

After the selection of a source of supply, both the buyer and the seller should regard their relationship as that of partners and should continue to negotiate or consider concessions which can be made.

The buyer should seek to build up a harmonious relationship with his source of supply by visiting the foundry and become known to as many people as possible at all levels with a view to furthering the relationship. The foundry should be regarded as if it were another department with both parties exchanging information in their mutual interest.

If the buyer builds a close relationship with the foundry, it is important to have a policy of using few suppliers and not send inquiries to every foundry that solicits business.

There may be cost savings from new sources and the buyer should keep in mind that changing circumstances can affect purchasing decisions. There are hidden costs and hazards in changing sources. The costs of expiditing and monitoring quality can increase in direct proportion to the number of suppliers. It is better, if the existing supplier is a good one, to negotiate and increase bargaining strength by concentrating the cost of a few, rather than many sources.

In cases where the buyer's company is forced to accept "penalty clauses" or is dependent on prompt delivery, there is perhaps an alternative to visits or telephone calls. First, obtain the supplier's agreement that delivery is part of the contract — that is, in return for a certain consideration (price) the supplier undertakes to supply good quality castings at the agreed date.

The advantage of this type of arrangement is that the top management at the foundry have a convenient means of measuring their delivery performance.

Some attempt should also be made to establish the cost of quality problems. In all probability, the cost of a rejection over a very broad spectrum of castings procurement is as much as five times the original price. Thus, if a batch of castings costing $1 each has a reject rate of 5%, the true cost of that batch is: five at $5 each plus 94 at $1 each, a total of $120 or $1.20 each, an increase of 20%.

The actual cost of quality or late delivery is an important consideration, the fact is that these costs may be significant and emphasize the need for a better understanding between buyer and supplier.

7.3 FOUNDRY COSTS

Because of the many types of foundries, the range of metals cast, and the wide range of casting processes, it is difficult to offer guidance on the subject. There are a few considerations which may be helpful.

Irrespective of the type of foundry, costs will break down into three main categories: materials, direct labor, and expenses. Materials may be subdivided into raw materials — pig iron, scrap and alloying constituents — and process materials — coke, sand, molding materials, bonding agents, shot, paint, etc.

Increasing mechanization and automation have made some impact on the foundry industry, but labor remains a high percentage of costs. Expenses are extremely variable depending on the facilities of the foundry and the method of costing.

There is no magic formula which can be universally applied to foundry costs, it is possible for the buyer to develop some idea of the costs for any foundry or particular casting.

For example, the cost of pig iron and alloying constituents is published information and most buyers will be aware of the level of scrap prices. it is usually possible to obtain some indication of the percentage ratio of pig iron, scrap, etc. in a typical melt charge. Having established an aproximate cost per ton of foundry output, based on the average cost per pound of a range of castings purchased, it is possible to arrive at a rough idea of the range of costs of the various materials, expressed as a percentage of total costs.

An idea of labor costs as a percentage of total cost can be calculated using average cost per ton of output related to the numbers of direct and indirect personnel and the assumed labor rate.

By multiplying the number of impressions on a pattern by the unit price for the casting, it is possible to estimate the return per flask expected by the foundry. Care must be taken to differentiate between cored and uncored casting, grade of material, molding method, and flask size when using this method of calculation.

Assuming castings represent a sufficiently large area of expenditure, the buyer would be advised to analyze casting prices by foundry process, grade of metal, size/weight, cored or uncored, molding flask size, and molding machine used. This is useful in comparing sources of supply and, related to quality statistics, provides a helpful guide to policy making and a useful tool in the negotiating process.

7.4 COST REDUCTION AND MUTUAL PROSPERITY

This is the basis of good casting procurement. There is no point in buying low cost castings if the supplier goes out of business. There has been a tremendous reduction in the number of foundries in the U.S. and the industry is still in decline. There are many reasons: environmental controls, the need for more investment, overseas competition from lower wage countries, etc. If buyers wish to continue purchasing in the U.S. and retain this important basic industry, there must be more professionalism, less secrecy, and more honesty regarding capability on the part of the foundry and a general recognition that mutual prosperity is essential.

Some actions on this subject which may influence buying policy include:

Action by foundries

- Increasing specialization
- Refusal to produce unsuitable/uneconomic castings
- Mergers into large groups
- Development of own machining facilities
- Improved technical liaison
- Assistance with preferred material specifications
- Improved production control arrangements

Action by buyers

- Design to use supplier flask sizes
- Encourage foundries to provide a design advisory service
- Better designs — more consultation
- Careful choice of process
- Closer quality monitoring and feedback

Action by both

- Use of optimum batch quantities
- Closer cooperation between buyer and seller
- Better planning — more reliable deliveries to achieve lower investment in inventories
- Long-term customer commitments to encourage investment on special plan
- Capacity planning — advance notice from buyer
- Steady load to optimize manufacturing costs
- Cooperation between buyer and seller to deal with monopoly suppliers with regard to scrap prices, availability, etc., and with representatives of government departments concerning licensing, regulations, etc.
- Introduction of computers for costing, production planning, and all administration

7.5 MACHINED CASTINGS

A small number of U.S. foundries have a machine shop. Putting aside possible constraints and assuming the buyer has complete freedom of action, consideration should be given to the purchase of machined castings if:

1. There is a high hazard and repairs may be needed.
2. The material has unusual properties requiring specialized machining techniques.
3. Castings have to be subjected to inter-operational testing or cleaning using specialized equipment.
4. There is insufficient capacity at customer's own works.
5. Transport costs of raw castings are prohibitive.
6. Castings are being purchased from overseas.

Usually the decision is fairly straightforward, but few U.S. foundries have set out to attract large volume orders for machined castings. The result has been that overseas foundries have made a strong impact. They have realized that in order to capture the business from overseas a large volume machining capability is essential.

Some foundries have a local subcontract machinist for small quantity work, but delivery can be a problem. It is necessary to have administrative capability to monitor the work satisfactorily.

The buyer seeking to purchase machined castings in quantity must consider the possibility of obtaining supplies outside the U.S. At the same time, the buyer should encourage U.S. sources to meet the need. Excessive overseas sourcing will lead to further decline in our foundry industry.

7.6 SUPPLIER SERVICES

Earlier, the need for buyer and seller to establish long-term relationships, involving mutual trust, and the encouragement of partnership was noted. The need to avoid fragmenting expenditure on castings was also stressed.

Progressive foundries are willing to provide advice on casting design to reduce cost and improve quality. The buyer has the right to expect this kind of assistance but in return the foundry will wish to have the opportunity of recovering the cost of this service in the volume of sales.

Services to be sought by the buyer include:

1. Practical advice on casting design, choice of material, and foundry process. In view of impending product liability legislation, the buyer should carefully examine the supplier's recom-

mendations and officially *endorse* them. They should be included in the purchase specification. Otherwise the foundry may not be willing to advise.

2. Regular monitoring of quality on customer's premises.
3. Regular commercial contact — early warning of impending difficulties, raw material price increases or decreases likely to affect casting prices, etc.
4. Constant updating of lead times and assist with urgent delivery requirements.

The buyer must ensure that the choice of a foundry will permit complete objectivity when advice or help is being sought.

It may be that objective advice is only possible when the foundry has a broad capability offering a range of processes and materials. This may not necessarily be the case and requires that the buyer exercise judgment. Fair play coupled with a competitive atmosphere and tough negotiation will identify those who have the ability to meet the qualtiy and price criteria.

7.7 BUYING FROM OVERSEAS

The United States has a very open economy and increasingly overseas sources are penetrating the U.S. market.

Some larger companies have already established considerable business outside the U.S., and castings buyers will at some stage be tempted to purchase overseas. This will bring a new set of problems. First, the overseas supplier will insist on strict observance of the contract. The buyer must be precise in specifying exactly what is required and the terms under which it is to be supplied.

While undoubtedly competent and economical suppliers can be found in a wide range of countries, any buyer not familiar with foreign markets should obtain expert advice about different trade practices, legal requirements, etc. While there is a host of different aspects to be considered, the following points are major aspects to ascertain that a fair comparison to local vendors is being made:

- *Purchase Price*

  In many cases, overseas vendors will offer in their native currency. It is important to consider the fact that currencies are subject to constant fluctuations which can be of significant magnitude. If, therefore, a purchasing contract is made on the basis of a foreign currency, the buyer accepts the exchange rate risk, i.e., the price in U.S. dollars at time of payment can be significantly higher or lower (whatever the case may be) than had been assumed at time of awarding the order.

Even if the price in the original quotation is stated in U.S. currency, it is important to carefully review the "small print" (conditions of sale). Quite frequently, offshore suppliers will quote in U.S. dollars, but stipulate in the conditions of sale, which form part of their offer, that the quoted price is subject to the exchange rate in effect at the time of quotation (or some other stipulated time) and subject to adjustment if changes occur between time of quotation and delivery. In essence then, a purchaser assumes the exchange rate risk just the same as if the price would have been quoted in foreign currency from the outset. It must be understood that this is common practice and not meant to mislead a buyer, but rather to provide the price, converted into U.S. dollars at a given exchange rate.

- *Product Liability*
  Product liability in the U.S. is quite strict and generally much more severe than in any other major market in the world. This has been recognized by all suppliers of goods and services by maintaining appropriate insurance coverage. In case of a major law suit under product liability legislation, the defendant will generally be supported by the supplier of a component, which led to the suit, as co-defendent, for example, the foundry. Most offshore suppliers, operating under their respective legislation, have either no insurance covering this risk or to a much lesser extent than would be the case in the U.S. Buying from foreign sources then means that the purchaser of such components or equipment will assume the respective risk himself without any or only limited recourse in the case of a law suit.

- *Governing Law*
  Practically all contracts with overseas suppliers stipulate the governing law, generally that in the country of the seller. In case of any legal dispute, the buyer will then be forced to pursue his rights in a foreign court, in a foreign language, and generally under conditions which are less favorable than what would be available to him in the U.S.

- *Freight Clauses*
  International trade is clearly ruled and governed by the Incoterms 1984, which are the generally accepted terms in world trade. Detailed definitions are contained in these rules with respect to the responsibilities of both parties. It must be noted that U.S. terms may, for example, use the same abbreviation, for example FOB, but with totally different meaning. Under Incoterms 1984, it is impossible to talk, for example, about a delivery "FOB our plant." Rather, FOB (free on board) is possible only when ship-

ment is made aboard a ship, whereby the buyer must reserve shipping space at the appropriate time to meet his vendors' delivery and questions such as who pays what are clearly regulated. For example, the risk of transport is borne by the buyer. He must insure, must pay for the freight to a North American port, take care of unloading in the port (including potential demurrage charges, import duties, etc.) All this is not to say that sourcing overseas should be avoided. But it should be done with the different risks clearly identified and should be performed only by those fully familiar with the rules and regulations of international trade to ensure that a proper comparison of the value (not necessarily the price) of each quotation has been made.

Other Considerations

Be sure of the position with regard to rejects and how long after delivery credit for defective castings is to be made. Define responsiblity for transportation of rejects. Reach agreement as to whether reject castings may be scrapped, and if so who will inspect.

If there is an obvious flaw in the design of the castings it is wise to insist on nondestructive testing or to purchase fully machined castings.

It is important when comparing quotations that shipping and other known costs have been considered. Before placing business overseas consider the cost of monitoring. Telephone conversations when another language is involved can be lengthy and frustrating. Remember when something has gone wrong it is comparatively easy for the supplier to excuse his failure or to avoid the problem simply by saying that he does not understand.

Importing castings into the U.S. is not difficult because shipping or air freight agents will be available with helpful advice. Generally speaking they are efficient and reliable and they only prosper from satisfied customers.

If in doubt consult other buyers known to be importing castings. Ask the supplier to give you names of other customers and contacts. Buyers are usually helpful to other members of the fraternity and it is good buying practice to take advantage of any information that is available.

7.8 QUESTIONNAIRES FOR FOUNDRIES

On the following pages you will find sample forms — one short and the other considerably more detailed — for foundry questionnaires. Casting buyers should amend these forms as necessary for their own operations.

Questionnaires for Foundries

Short Form

Name of company ______________________ Name of parent company ______________________

Address ______________________ Address ______________________

Tel/Telex nos. ______________________ Tel/Telex nos. ______________________

Number of employees ______________________ Number of employees ______________________

Grades of material produced ______________________

Annual tonnage output each grade ______________________

Molding processes ______________________ Process used for making cores ______________________

Types of melting facilities ______________________

Patternmaking facilities ______________________

Machining facilities ______________________

Percentage capacity own use ______________________

Preferred quantities ______________________

Weight range ______________________

Flask size ______________________

Specialty castings ______________________

Quality approvals ______________________

Industries served ______________________

Principal areas currently served ______________________________

Percentage output exported ______________________________

Future plans/development and projected dates ______________________________

____

Long Form

Name of company	Name of parent company
______________________	______________________
Address ______________________	Address ______________________
______________________	______________________
Tel/Telex nos. ______________	Tel/Telex nos. ______________
Number of employees __________	Number of employees __________

Grades of material produced/process used/tonnage (see example)

Grade	Molding process	Average weekly tonnage	Maximum weekly tonnage

Types of melting facilities

Electric induction/electric arc cupola/other — please state

____

Patternmaking facilities ☐ yes ☐ no Number of employees __________

Machining facilities ☐ yes ☐ no Number of employees __________

Type of machines: Conventional/NC/CNC/lathes/borers vertical or horizontal/drills, bench, radial, multi-spindle/other — please state. Delete as appropriate.

Is pattern/machine shop directly controlled by foundry management? ☐ yes ☐ no

Is above capacity tied to own use/associate/group companies? ☐ yes ☐ no

If so, what percentage/tonnage? ______________________

Preferred quantities/grades

Grade	Quantity preferred	Weight range	Box Size	Maximum daily molds cored	uncored

State if any loose molding prototype facilities available.

____

Coremaking facilities

Oil sand/CO_2/Shell/No-bake, chemical bonded, other — please state

____

Specialty castings

State type/industry/process, etc. ______________________________

____

Quality approvals — MoD, etc. ______________________________

State number of personnel directly and exclusively engaged on quality control ______

State type of inspection facilities and give details of nondestructive testing equipment

____

Industries currently served ______________________________

Own transportation ______________________________

Principal areas currently served ______________________________

Percentage output exported ______________________________

Future plans/development and projected dates ______________________________

____

Example

Grades

Gray iron/Meehanite/G10 (150) G12 (200) G14 (220) G17 (260)

Steel	Carbon	Low alloy	Stainless	Whole range
Aluminum Zinc Copper alloys		Whole range of specs		

Process

Greensand	CO_2	No-bake	Shell	Lost Wax
Shaw	Pressure diecast		Gravity diecast, etc.	

8

Conditions for Casting Purchases

8.1 FOUNDRY CONDITIONS OF SALE

The more important points normally covered by the foundry conditions of sale are as follows:

1. The foundry reserves the right to amend prices if the quotation and/or order is not accompanied by full and correct information.
2. Alterations to design, weight, specification, or quantities may involve adjustments to price.
3. Responsibility will not be accepted for the accuracy of castings produced from patterns supplied by the buyer.
4. The buyer will be responsible for maintaining patterns in good repair.
5. A charge will be made for mounting the customer's patterns onto the supplier's molding equipment.
6. In some circumstances, a charge will be made for marking out sample castings.
7. Insurance of patterns and gages may be the responsibility of the buyer.
8. Transportation by rail or trucking is often paid by the foundry but special transportation arrangements are almost invariably charged to the buyer.
9. Credit may be declined for castings rejected more than six months from the date of delivery.
10. The buyer will normally not be compensated for machining losses due to defective castings.
11. The foundry may return to the buyer for safe custody all patterns from which castings have not been ordered for a period of 3–5 years. (Some encourage return after 2 years.)
12. Written approval of samples and specific instructions to proceed

may be required before production commences.

13. Unless otherwise agreed, the foundry may claim that specifications relating to chemical and physical tests do not constitute a guarantee.

14. Claims for free replacement of castings will not be entertained in respect of castings found to be defective as a result of faulty design or improper construction of patterns supplied by the buyer.

15. The foundry will disclaim any liability for damages arising from infringements of patents while working to the buyer's instructions.

The buyer is advised to study these points and negotiate these points and negotiate alternative conditions where those proposed by the foundry are unacceptable.

The buyer should insist that all repairs and alterations to patterns require a quotation and appropriate authorization before any chargeable work is completed.

Where strict adherence to delivery date is important and the foundry is unwilling to accept a penalty clause, an alternative arrangement would be considered.

If guaranteed margins of chemical and physicial properties are required this should be stated at the outset and included in the buyer's conditions of purchase on the face of the purchase order.

The European Foundries Conditions of Sale cover much the same points as those normally imposed by U.S. suppliers. The main difference is that the European conditions are much more comprehensively written up.

For comparison, shown on page 60 through 63 is a copy of the "Conditions of Sale" of the Council of Iron Foundry Associations and of the Committee of European Foundry Association.

8.2 TERMS OF PAYMENT

Failure to observe contract terms with regard to payment is one of the industry's major problems. A contract to buy and sell respectively has to have the following elements:

1. An offer made and accepted
2. Mutual intent to enter into a contract
3. Consideration
4. Capacity to contract
5. Lawful purchase

As, for example, stated in D.L. Marston's work: "Consideration is an essential part of an enforceable contract. As Black's Law Dictionary points out, it is the cause, motive, price, or impelling influence that induces a contracting party to enter into a contract."

Specifically, in this world of "just in time," delivery time is as important as price. Not to deliver on time is a clear breach of contract and in the end as unethical as not meeting the agreed upon time for payment.

The company which persistently delays payment to suppliers in breach of contract arrangements has to pay the cost, in terms of later deliveries and, eventually, higher prices. Companies habitually taking such advantage of their suppliers should conduct a careful analysis of the results of this practice. Late deliveries cost money in many ways: overtime work to make up lost time, telephone calls and visits to chase delivery, which results in inflated prices. Unfortunately it is the buyer, rather than the company's financial officers, who is likey to receive the blame for poor supplier performance and who may lose credibility with the supplier.

In general, foundries suffer when their customers pay late. They cannot delay payments to their suppliers.

8.3 REJECTIONS

The discovery of a batch of faulty castings which have exceeded the maximum time interval between delivery and rejection is a problem of the casting user. The buyer should observe the following rules.

1. Clarify with the supplier any special circumstances which will necessitate an extension of this time interval.
2. Endeavor to use castings as soon as possible after delivery.
3. Use stocks in strict rotation and report any problems immediately.

8.4 MACHINING LOSSSES

Users of castings may at some stage consider charging the foundry with machining losses due to faulty castings. Foundries consistently try to avoid this charge.

The principal objection to charging these losses to the foundry is the addition of another factor into the cost equation. This would mean that a percentage would be added to cover this contingency and the risks may be overstated.

The answer to the problem is to concentrate on quality improvement. Prevention is always better than cure and in this case, avoiding the problem actually reduces costs.

CONDITIONS OF SALE

1. **Orders**

Any order based on this tender is subject to the following conditions. This tender is open for acceptance (unless previously withdrawn) for not more than ______ days from the date thereof unless otherwise stated. It is subject to confirmation on receipt of order.

2. **Prices**

(a) Unless the order is accompanied by sufficient information, drawings, and patterns to enable work to proceed forthwith, the seller is at liberty to amend prices quoted herein to cover any increase in costs during the period of delay caused by the lack of such details.

(b) Any alterations by the buyer in design, weight, quantities, or specification and any suspension of work due to instructions or lack of instructions will involve adjustment of the agreed or quoted prices, if the costs are affected thereby.

(c) Prices quoted for unmachined castings unless otherwise stated.

3. **Terms**

(a) Prices quoted are net. Accounts are due for payment not later than end of the month following (the month of) dispatch. When deliveries are spread over a period, each consignment shall be invoiced when dispatched and each month's invoices shall be treated as a separate account and payable accordingly.

(b) Should the buyer cancel, suspend, or reduce a quantity requirement, including cancellation, suspension, or reduction of "firm" schedules, then any work already in progress will be delivered and invoiced in accordance with the earlier instruction of the buyer and shall be paid for by the buyer.

(c) If and so far as work upon castings has been necessarily commenced in advance of "firm schedules" in order to provide, in accordance with a normal process time cycle, for the deliveries indicated by a "tentative schedule," the buyer shall be liable to accept delivery of those castings at the times and in quantities so indicated and to pay therefor.

4. **Patterns**

(a) Where the buyer supplies patterns the quotations of the seller assume that such patterns are in good condition, true to drawings, and entirely suitable for the seller's methods of production and for the production of the castings in the quantitites required.

(b) For mutual benefit, when new patterns or equipment are to be made, the seller requires to be consulted.

(c) Replacement of and alterations or repairs to buyers' patterns, or equipment due to normal wear and tear shall be paid for by the buyer.

(d) Where patterns are not supplied by the buyer, only such patterns as are specially made and separately charged in full shall become the property of the buyer when paid for.

(e) Carriage on patterns and equipment supplied by the buyer will be paid by the seller in one direction only.

(f) The seller takes all reasonable care to protect buyers' patterns while they are on the seller's premises but does not accept liability for any loss, damage, or expense arising from any cause whatsoever which does not directly and solely result from a failure by the seller to exercise such reasonable skill and care.

(g) The buyer shall be responsible for the custody of his patterns from which no castings have been ordered for a period of three years.

5. **Transportation**

Unless otherwise specified, prices quoted include delivery to destinations and, at the option of the seller, may be either:

(a) By rail transportation to the address stated on the buyer's inquiry or an equivalent address.

(b) By truck transportation on suitable roads to the address stated on the buyer's inquiry.

Extra cost of special delivery, at buyer's request, by passenger train or other express methods will be charged to buyers.

6. **Delivery**

Time for delivery is estimated as accurately as possible, but is subject to any delays or breakdowns beyond the control of the seller and is not guaranteed.

The period specified for delivery on the seller's quotation:

(a) Is exclusive of any period occupied in making, altering, or adapting patterns or in any experimental work connected with the castings.

(b) Shall commence only after the receipt of written instructions to proceed together with all necessary information, drawings, and (if to be supplied by the buyer) patterns or equipment.

(c) Shall (if a sample casting is to be submitted for buyer's approval) commence only from date of receipt of written approval.

(d) Owing to the difficulty of producing exact quantities of castings the seller reserves the right to deliver up to ______ %, in excess of the quantities ordered unless special agreement has been made to the contrary.

7. **Damage, Shortage, or Loss**

The seller does not accept responsibility for any damage, shortage, or loss in transit unless:

(a) Damage or shortage is notified in writing both to the seller and to the carriers within three days of receipt of goods and the goods have been signed for as "not examined" and have been handled by the buyer in accordance with Carriers' conditions, or

(b) Nondelivery (in the case of total loss) is notified both to the seller and to the carriers within the carriers' permitted period.

8. **Samples**

Samples submitted will be payable by the buyer unless returned to the seller's works, carriage paid, within one month from the date of dispatch.

In all instances where the seller is working from a new pattern, an altered pattern, or a pattern fresh to the seller's foundry, the seller may submit sample castings for approval before executing the bulk of the order, which will only be commenced on receipt of such approval in writing.

Where small quantities only are required submission of samples will be made only if such is requested by the buyer at the time of placing the order.

9. **Tests**

Unless otherwise stated, the cost of supplying, machining, or testing all test pieces required by the buyer will be charged extra.

When figures or particulars relating to physical or chemical properties are indicated, they are to be regarded as a general guide only, and constitute no guarantee from seller unless specified margins have been agreed at the time of placing the order.

While every effort is made to provide sound castings no express or implied warranty is given by the seller as to the fitness or suitability of castings for any particular purpose whether such purpose is known to the seller or not.

10. **Defects**

(a) The invoice value of any castings made by the seller and proved to be defective in workmanship or materials will be credited to the buyer, provided that the castings are returned to the seller within ______ months from date of dispatch. Any such agreed defective castings will be replaced (and re-invoiced) at the price credited or made serviceable for their original purpose free of charge.

(b) The buyer shall make every effort to ascertain any possible defects as soon as possible after delivery of the castings, including any necessary tests or inspection during or after machining. Immediately after discovery of any such defects or alleged defects, the buyer shall notify the seller in writing and give the seller a reasonable opportunity to take prompt measures to prevent a repetition of the defect.

(c) Defective castings will not form the subject of any claim for labor, machining costs, or other expenditure theron or for resultant loss or damage arising out of any such defect.

(d) Expenditure by the buyer on the salvaging of defective castings may be a matter for agreement between buyer and seller, but in the absence of such agreement it shall not be chargeable to the seller and any such salvaging operation shall not be proceeded within any manner liable to prejudice the opportinity of the seller to take the earliest possible steps to avoid a repetition of the defect in any further castings he may be making.

(e) No claim for free replacement or otherwise will be accepted in respect of any castings found to be defective through faults in the design or construction of patterns supplied by the buyer.

11. **Head Metal**

Any head cast integral with the casting to ensure soundness and delivered to the buyer will be credited at the full casting rate if returned to the seller within six weeks from date of dispatch.

12. **Packing**

Unless otherwise specified, packing cases and packing materials will be charged extra, but will be credited in full on return to the seller's works and in good condtiion within one month of receipt by the buyer.

13. **Export**

Contracts for export will be subject to separate Condition of Sale.

14. **Infringements**

The buyer shall indemnify the seller against all damages, penalties, costs, and expenses arising out of and loss suffered as a result of the infringement of any Patent Registered Design, Trade Mark, Trade Name, or copyright or any claim for such infringement or any claim to passing off involved in or arising out of work carried out in accordance with the buyer's specification.

9
Purchase Records for Casting Buyers

9.1 PURCHASE HISTORY CARD

This card should carry the following essential information:

1. Part number.
2. Drawing number.
3. Description.
4. Weight.
5. Standard material specification.
6. Supplier(s).
7. Special notes relating to quality or inspection standards. This can be achieved by sending the supplier suitably coded quality control instructions. Care should be exercised in developing this system to avoid misinterpretation.
8. General size description — assist in determining number of castings per mold.

The purchase order should always use the same format. If purchase orders are prepared by computer, provision must be made in a meaningful code to ensure that the item is specified precisely and that the supplier fully understands the method of coding.

9.2 MASTER PATTERN INDEX

Whatever the expenditure on castings, it is essential to have a record of the pattern equipment available and the following information should be recorded for all castings.

1. Part number.
2. Drawing number.
3. Description.
4. Supplier(s).

5. Type of pattern — wood, metal, plastic, etc.
6. Number of impressions on plate.
7. Number of coreboxes.
8. Molding flask size.
9. Flask pin centers.
10. Type of molding machine and process.
11. Cost of pattern
12. Date of manufacture and manufacturer.
13. Repairs cost and date.
14. Movement dates transferred to other foundries.
15. Other information such as:
    - Total number of castings produced
    - Amortized to date
    - Pattern cost

10
The Quotation

The buyer is best served by the concentration of expenditure on a few rather than many sources. Using available information, the buyer should request quotations from those foundries with whom he is prepared to do business in the event to their submitting the most attractive bid (i.e., best offer in terms of price, quality or delivery, or any combination of these and other factors which may influence the purchasing decision).

Inviting quotations from poorly equipped low cost producers, whose ability to meet required standards may be in doubt, and setting their prices against those sources known to observe higher standards as a negotiating ploy, leads only to disaster.

The buyer must beware of the foundry which, as a short-term expedient, is prepared to undercut all competition. Such foundries, in the final analysis, discard unattractive items, usually at times inconvenient to the buyer, and what appears to be a saving becomes a significant loss.

To attract the best quotation the information given to potential sources must be comprehensive and should include the following details.

1. Part number.
2. Drawing number, and the number of copies of drawings required.
3. Description.
4. Material specification, U.S. standard or equivalent.
5. Annual requirement broken down into a monthly or weekly schedule.
6. Batch quantities required and commencing date.
7. Pattern equipment — if existing give details, if new pattern required, request prices and details, state for prototype or production use.
8. Actual or estimated casting weight.
9. Special features required such as:
   - Finish
   - Cleaning

- Heat treatment
- Nondestructive tests
- Fin tolerance
- Clearances for a jointing component

10. Service conditions:

- Abrasion
- Corrosion
- Temperature

The drawing accompanying the quotation should indicate the following standard material specification or mechanical properties desired:

- Tensile
- Hardness
- Dimensional tolerances
- Maximum/minimum temperature characteristics
- Pressure tightness (if important)
- Machining data
- Location of part numbers, trade marks, machine locators
- Gaging points (if applicable)

Require separate prices for castings and pattern equipment. Drawings used for quotations should be marked "for quotation purposes only."

Note
Having provided the potential suppliers with comprehensive details of the requirement, insist that quotations are complete and do not produce more questions than answers.

If the requirement is for intricate castings, the quotation should specify those features which cannot be achieved in the "as-cast" condition.

Quotations should be accompanied by a marked drawing showing the precise form of supply and identifying any special machining allowances.

Ambiguities regarding price, quality, material specification, delivery, nondestructive testing, terms of payment, etc., should be identified and eliminated before any quotation is considered.

See also the sample request-for-quotation (RFQ) form in Chapter 17, which can be used as a checklist for all pertinent information for the foundry.

11

The Purchase Order

The purchase order is an important legal document and frequently provides the minimum information to set in motion a complex chain of events. Often the outcome is achieved by extensive use of the telephone and visits to and from the supplier. Sooner or later these costs have to be recovered and the inefficiency paid for in a variety of ways.

Most buyers are familiar with stories of 100 tins of paint becoming 100 tons as a result of a small typographical error. Fewer may be aware that "SG Iron" has been wrongly interpreted as "Soft Gray Iron" with disastrous results. These are possibly extreme examples of the consequences of carelessness; nevertheless, they illustrate the point that accuracy is important and ambiguity should be avoided.

A great deal of information on castings specifications is available and designers and buyers are advised to ascertain the appropriate standard specification for the grade of material they require. This is preferable to supplier brand names or metal alloy reference numbers.

Most of the preliminary work necessary for successful castings procurement has already been discussed and all that remains is to finalize the arrangements with the chosen supplier. The procedure for the purchase of castings may be as follows.

1. Ascertain that design and production engineering departments are satisfied that all technical points have been cleared and that the production drawing has been amended to show the agreements reached on such points as:
   - Machining allowances
   - Machining data
   - Machining locator points
   - Position of cast part number or other foundry identification marks
2. Make a final check on supplier's conditions of sale.
3. Prepare the purchase order giving:
   (a) The full material specification
   (b) Casting pattern or drawing number
   (c) Total number of castings required (if applicable)

(d) Date for delivery of samples
(e) Batch quantities
(f) Schedule of delivery
(g) Quoted (or negotiated) price
(h) Quotation date and reference (or details of price agreement)
(i) Details of pattern equipment to be supplied by or to the supplier
(j) Price of pattern equipment (if applicable)
(k) Details of nondestructive testing required
(l) Definition of responsibility for marking of samples
(m) Details of any special requirements or service conditions (even if noted on drawing)
(n) Terms of payment (settlement discounts or as applicable)

Take care to enclose a certified copy of the drawings for production purposes and ensure that there is a reference to this enclosure on the purchase order.

12
Casting Delivery Reliability

12.1 EXPEDITING

The requirement for expediting activity may be a reflection of the purchasing policy and may be due to one or several of the following reasons.

1. Bad choice of supplier.
2. Poor payments.
3. Incorrect pricing.
4. Inadequate lead times.
5. Unsuitable pattern equipment.
6. Unrealistic scheduling.

Supplier election is a key factor and the buyer should visit each foundry with whom business is intended. He should look for good standards of housekeeping, positive management attitudes, production control facilities and methods, process controls, and quality assurance procedures.

The buyer should assess the work rate in the foundry observing casting flow through the finishing department and into the dispatch area. He should examine castings waiting for dispatch and compare them with his own requirements of shape, size, and finish.

The buyer must be satisfied on all these points and should not be tempted by price advantages. Suppliers with inadequate management resources, skill, and experience in the production of the required type of castings will invariably fail on delivery.

Poor payment is the surest way to encourage unreliable delivery performance. Priority will always be given to work which is rewarded with payment on the due date. If extended credit is needed during difficult times, the buyer should think carefully about the cost of this credit.

Incorrect pricing, whether due to unfair competition, tough

negotiation, or simple errors, will lead to delivery failure, for obvious reasons.

If lead times are inadequate, delivery reliability will only be achieved at enormous cost. Casting users would do well to concentrate their attention within their own organization to reduce the incidence of urgent requirements to the absolute minimum.

If the pattern equipment is inadequate — insufficient number of impressions, poor state of repair, or not of the type required by the foundry — the result is likely to be unreliable deliveries coupled with bad quality and unfavorable prices.

A frequent cause of delivery failure can be traced to transportation problems. Not all foundries have their own trucking facilities. Attempts made to optimize transportation costs may cause delay while a suitable load is collected together for a particular destination. Transportation arrangements are seldom considered in negotiations with the foundry, but should be carefully analyzed when delivery reliability may be a problem.

12.2 EXPECT SUPPLIER COOPERATION

Suppliers should be expected to cooperate with buyers to achieve best possible results. Depending on the size of the account it may be reasonable that they should allocate one staff member to follow that account and to make regular visits. Excellent results can be achieved through a consistent program using this procedure.

The foundry should acknowledge all orders and comment on delivery prospects. Similarly, buyers have a right to expect reliable delivery promises to be made against all schedules. Failure to comment on delivery prospects should imply the supplier's agreement of meeting the requirement. The buyer should make a point of the importance of delivery in the negotiating stage whether this be before placing individual orders or renewing/reviewing annual contracts.

Capacity planning on a daily, weekly, or monthly basis will avoid many delivery failures. If the buyer has a sizable account, it is reasonable that the supplier should be prepared to cooperate by allocating the buyer a certain capacity and a degree of flexibility. This requires fair play on the part of the buyer.

12.3 SCHEDULING REQUIREMENTS

Unrealistic scheduling — whether quantities are too small, too large, or too frequent, with regard to the pattern equipment in use — inevitably means delivery performance will not be reliable when care is not taken to match the equipment to the requirements which are realistic in terms of foundry capability.

Agreements made between buyer and seller with regard to lead times for schedule changes should be honored by both parties. The buyer should, if possible, avoid rapid changes in schedules. It is a valid complaint on the part of foundries that buyers overschedule their requirements until excessive stock accumulates and then a sudden drop in volume or a complete cancellation of a schedule occurs.

Suppliers of castings do not expect a perfect schedule, with no changes, but equally buyers, or their production managements, should recognize the difficulty for foundries to plan order out of chaos. It may be desirable that the customer is always right; however, it is unrealistic to expect this to be translated into practical results.

12.4 VISITS TO AND FROM SUPPLIERS

Visits to the foundry ensure that the buyer will be known to the appropriate personnel which is important to ensure delivery arrangements. Buyers should strive to create conditions which suppliers will be sending production control or sales personnel to monitor customers' delivery requirements at regular intervals. Visits to the foundry by the buyer should retain an element of surprise occasionally. Suppliers of castings tend to postpone actions which should have been taken very much earlier until near a routine visit.

Delivery reliability is achieved by consistent purchasing policy and constant vigilance coupled with intelligent planning and sensible demands upon the capacity available. It is seldom accomplished by bullying tactics from those users who make only modest demands on foundry capacity.

13

Quality/Inspection Assurance and Process Control

Close liaison between purchasing and quality assurance personnel is essential for successful castings buying. Both should be involved with the supplier with good communication between the two departments.

Visits to and from the foundry are encouraged and, by means of a formal or informal system, quality approval should be sought before new suppliers are introduced.

Every foundry may not have highly sophisticated inspection equipment or an army of white coated inspectors, but quality assurance procedures of the foundry should be satisfactory for the type of work they are required to produce.

13.1 STATISTICAL PROCESS CONTROL

Quality assurance programs using Statistical Process Control (SPC) are widely used in the foundry industry and should be reviewed to ensure the program does in fact control the processes intended. Statistical Process Control is a scientific method of analyzing data and using the analysis to control a process. It may be applied to any process which generates numbers and may be used in any field of activity. An example may be the plotting of numbers on a process control chart regarding the moisture content of sand which will assist in establishing a specified moisture content range for the operating personnel to establish manufacturing consistency. When the moisture falls outside of the SPC specifications, scrap and a rough surface on the casting results.

Specifications should be realistic for the process being used and the casting being produced.

A brief reference to techniques and terminologies follows.

Brinell hardness A comparative measurement of hardness obtained by pressing a steel ball of known diameter into the surface of a material under a standard pressure and measuring the diameter of the impression. The higher the number the greater the Brinnell Hardness (BHN) (see also Rockwell Hardness).

Charpy test A specially prepared bar fixed at both ends is struck by a pendulum type of device. The energy absorbed in fracturing the specimen is measured by the height to which the pendulum rises (see also Izod test).

Dye penetrant test A nondestructive method of examining castings for *surface* defects and cracks. A dye (usually red) is sprayed on the area to be tested. After allowing the dye to remain for a prescribed time, a solvent cleaner is sprayed on the area and the dye is wiped clean. A developer (usually white) is then sprayed on the area. A crack or surface defect will then be observed by a bleeding through "indication" on the white developer background.

Gaging Enables castings to be checked on critical dimensions on a production basis.

Izod Test The method is similar to the Charpy Test except that the specially prepared test bar is supported on one end and struck at the opposite end by a pendulum type device.

Magnetic particle inspection Suitable for use only on ferrous magnetic materials. A powdered magnetic material is applied to a magnetized surface. Cracks are readily identified by the accumulation of iron powder at the flux leakage resulting from the crack. Detection can be facilitated by pre-painting the area white or using iron powder coated with fluorescent material.

Pressure test Apertures are sealed off and the casting filled with water or oil. Predetermined pressures are built up and maintained while the casting is examined for leaks.

Proof machining Limited machining operations designed to expose faults at an early stage.

Radiography A non-destructive test using X-Rays as a source for detecting the internal soundness and integrity of castings.

Rockwell Hardness Similar to Brinell Hardness in that hardness of the material is determined by indenting with a steel ball (B scale) or diamond cone (C scale) under specific loads.

Spectrographic analysis A means of examining suitable machined or ground test pieces and verifying the existence of known elements in correct proportions.

Surface roughness comparator A visual means of evaluating the cast surface of a casting. A surface roughness comparator identifies various grades of roughness of a casting. A casting displaying several grades of casting surface roughness having assigned values ("C" or "RMS") used for visual comparison and evaluation of as cast surfaces of a casting.

Tensile test Test bars cast from the same heat as the casting can be used to obtain an *indication* of the mechanical and physical properties of the casting. Results obtained may vary in different parts of the castings and must therefore be carefully interpreted.

Ultrasonic test Internal defects can be identified by transmitting high frequency waves through the material. Defects are identified by the reflections of the transmitted signal which are generated when a defect is encountered. The reflected signals are reproduced on an oscilloscope which enables the operator to establish the precise position of the defect.

Visual inspection A simple inexpensive procedure for eliminating castings possessing obvious faults.

14 Comments For Buyer Consideration

14.1 BRAND NAMES

Generally speaking, there is no magic in a brand name. The purpose is almost invariably to establish brand loyalty and the advice to castings buyers is: wherever possible use U.S. Standard or International Standard specifications.

14.2 CASTING WEIGHT DISCREPANCIES

While the castings buyer will find it useful to convert casting prices into price per pound, it is a mistake to purchase on this basis.

In the event of an agreement between buyer and seller to adjust prices up or down, depending upon actual versus estimated weight, the buyer must ensure that, if the weight of the casting shows an increase, the price is increased only by the metal content; labor and other costs remain unaltered.

It is advisable to agree in the quotation stage the details of any price adjustment clause. This provides for accurate comparisons between various suppliers' quotations.

14.3 PATTERN EQUIPMENT CONDITION

This is a neglected area resulting in quality and delivery failure and often increased cost. Suppliers should be encouraged to examine and report on pattern equipment condition at the end of each production run or at regular intervals. The condition should be recorded by the buyer.

Foundries are often guilty of returning pattern equipment to stores on completion of a production run without an examination. Pattern equipment will become the focus of attention when urgent delivery is required, castings have been rejected, or a price increase challenged.

Buyers are advised to budget annually for repairs and to insist that foundries provide the necessary information.

14.4 POOR PROCESS SELECTION

The foundry industry today offers a wide choice of processes. Basic information has already been given on this subject and care must be exercised if optimum results are to be achieved.

If in doubt consult the foundry offering a range of processes or seek independent advice. Poor process selection is responsible for many of the problems which casting users experience.

Assessment of casting quality requirements must be high on the list for the buyer when making his choice.

14.5 PRODUCT LIABILITY — THE FUTURE

It is certain that over the next few years there will be important developments in this field. This will demand more professionalism on the part of the buyer and it is hoped that at least some of the information contained in this guide will encourage a cautious approach to the castings buying problem. In so doing more information on the subject will be acquired, and buyers will be encouraged to make full use of the technical support foundries offer their customers.

Legislation on product liability will almost certainly lead to better castings buying practice.

15
Hints for the Buyer

The following is a list of helpful hints which have been noted under other headings.

1. Always give potential suppliers casting weights if known.
2. All quotations submitted by the foundry solely on a price per pound basis or any prices which appear to have been calculated in this manner should be carefully evaluated.
3. Insist on the supplier stating estimated weights on all quotations.
4. For the purpose of comparing suppliers, types of castings and differing metal specifications, and for statistical reasons, it is helpful to calculate the price per pound from actual or estimated weights, but *do not* expect dissimilar castings in shape, weight, or size to convert to a constant price per pound.
5. When changing sources, advise the new supplier of any quality problems. Do not transfer your quality problem with the pattern equipment, give him a complete description of the tooling.
6. Encourage suppliers to suggest design modifications to improve quality and to discuss their problems openly.
7. Always use the best pattern equipment you can afford relative to castings demand.
8. Seek to reduce costs by maximizing the number of castings per mold.
9. Encourage suppliers to report on pattern equipment periodically.
10. Establish responsibility for routine pattern repairs.
11. Obtain estimates annually for pattern repairs/renewals to facilitate preparation of budgets.
12. Involve the supplier in your problems and advise any machining changes or introduction of new machine tools as soon as possible.
13. Insist on routine monitoring visits from suppliers according to the volume of business transacted.

14. Do not overspecify.
15. Do not use vague specifications.
16. Do not expect good, consistent quality from cheap equipment or at an unrealistic cost.
17. Look for positive, enthusiastic management attitudes among your suppliers and try to cultivate long-term relationships.
18. Get to know who to contact for what purpose and encourage technical personnel to contact their opposite numbers at the foundry.

16
Glossary of Foundry Terms

AQL Acceptable Quality Level. A quality level established on a prearranged system of inspection using samples selected at random.

As-cast condition Casting without subsequent heat treatment.

Backing sand The bulk of the sand in the flask. The sand compacted on top of the facing sand which covers the pattern.

Binder The bonding agent used as an additive to mold or core sand to impart strength or plasticity in a "green" or dry state (also see green sand).

Burn-on sand Sand adhering to the surface of the casting which is extremely difficult to remove. May be due to soft molds, poor sand compaction, insufficient mold coating (graphite) paint, or high pouring temperature.

Chaplet A small metal insert or spacer used in molds to provide core support during the casting process. Correctly used will fuse with the molten metal.

Charge A given weight of metal introduced into the furnance.

Chill A metal insert in the sand mold used to produce local chilling and equalize rate of solidification throughout the casting.

Cleaning Removal of runners, risers, flash, surplus metal, and sand from a casting.

Cold shut A surface imperfection due to unsatisfactory fusion of metal. Caused by insufficient fluidity, low pouring temperature, improper choice of alloy, or inadequate runner systems.

Cope The top half of a horizontally parted mold.

Core A metal insert in a die to produce a hole in a pattern. (See *preformed ceramic core*.)

Core assembly An assembly made from a number of cores.

Corebox The wooden, metal, or plastic tool used to produce cores.

Coreprint A projection on a pattern which leaves an impression in the mold for supporting the core.

Core wash A suspension of a refractory material applied to cores and dried. (Intended to improve surface of casting.)

Crush Displacement of sand at mold joints.

Cupola A cylindrical straight shaft furnace usually lined with refractories, for melting metal in direct contact with coke by forcing air under pressure through openings near its base.

Cure To harden.

Die A metal form used as a permanent mold for die casting or lost wax process.

Dowel A pin of various types used in the parting surface of parted patterns or dies to asure correct registry.

Draft Taper on the vertical sides of a pattern or corebox which permits the core or sand mold to be removed without distorting or tearing of the sand.

Drag The bottom half of a horizontally parted mold.

Ejector pins Movable pins in pattern dies which help remove patterns from the die.

Facing sand The sand used to surround the pattern, which produces the surface in contact with the molten metal.

Feeder Sometimes referred to as a riser. A vertical channel in the mold (part of the runner system) which forms the reservoir of molten metal necessary to compensate for losses due to shrinkage as the metal solidifies.

Finish allowance The amount of stock left on the surface of a casting for machining.

Finish mark A symbol (f, f1, f2, etc.) appearing on the line of a drawing that represents the edge of the surface of the casting to be machined or otherwise finished.

Flask A metal or wood rigid frame without top and without fixed bottom used to hold the sand of which a mold is formed; usually consisting of two parts, cope and drag.

Foundry returns Metal in the form of gates, sprues, runners, risers, and scrapped castings of known composition returned to the furnance for remelting.

Gas porosity A condition existing in a casting caused by the trapping of gas in the molten metal, or by mold gases evolved during the pouring of the casting.

Gate (ingate) The portion of the runner where the molten metal enters the mold cavity. Other gates: bottom — any gating system by which metal enters the mold cavity at the bottom.

Green sand Moist clay bonded molding sand.

Heat A single furnace charge of metal to be used for pouring directly into mold cavities; a heat may be all or part of a master heat.

Heat treatment A combination of heating and cooling operations timed and applied to a metal or alloy in the solid state in a manner which will produce desired properties.

Hotbox process A furan resin-based process which uses heated metal coreboxes to produce cores.

Hot tear Irregularly shaped fracture in a casting resulting from stresses set up by steep thermal gradients within the casting during solidification and too much rigidity of the core or mold material.

Inclusions Particles of slag, refractory materials, sand or deoxidation products trapped in the casting during solidification.

Investment casting process A pattern casting process in which a wax or thermoplastic pattern is used. The pattern is invested (surrounded) by a refractory slurry. After the mold is dry, the pattern is melted or burned out of the mold cavity, and molten metal is poured into the resulting cavity.

Ladle A container for molten metal used to transfer metal from the furnace to the mold.

Locating pad A projection on a casting that helps maintain alignment of the casting for machining operations.

Locating surface A casting surface to be used as a basis for measurement in making secondary machining operations.

Master pattern The object from which a die can be made; generally a metal model of the part to be cast with process shrinkage added.

Mechanical properties Those properties of a material that reveal the elastic and inelastic reaction when force is applied, or that involve the relationship between stress and strain; for example, the modulus of elasticity, tensile strength, and fatigue limit. This term should not be used interchangeably with "physical properties."

Metal lot A master heat that has been approved for casting and given a sequential number by the foundry.

Mold Normally consists of a top and bottom form, made of sand, metal, or any other investment material which contains the cavity into which molten metal is poured to produce a casting of definite shape and outline.

Mold cavity The impression in a mold produced by removal of the pattern. It is filled with molten metal to form the casting. Gates and risers are not considered part of the modl cavity.

Mold coating Coating to prevent surface defects, i.e., metal penetration and improve casting finish (see core wash).

No-bake process Molds/cores produced with a resin-bonded air-setting sand. Also known as the air set process because molds are left to harden under normal atmospheric conditions.

Parting line The line showing the separation of the two halves of the mold.

Pattern The wood, metal, or plastic shape used to form the cavity in the sand. A pattern may consist of one or many impressions and would normally be mounted on a board or plate complete with a runner system.

Pattern draft The taper allowed on the vertical faces of a pattern to permit easy withdrawal of pattern from the mold or die.

Pattern layout Full-sized drawing of a pattern showing its arrangement and structural features.

Patternmaker's shrinkage The shrinkage allowance made on all patterns to compensate for the change in dimensions as the solidified casting cools in the mold from freezing temperature of the metal to room temperature. Pattern is made larger by the amount of shrinkage characteristic of the particular metal in the casting and the amount of resulting contraction to be encountered. Rules or scales are available for use.

Permeability	The property of a mold material to allow passage of gases.
Physical properties	Properties of matter such as density, electrical and thermal conductivity, expansion, and specific heat. This term should not be used interchangeably with "mechanical properties."
Pig iron	Blocks of iron to a known metal chemical analysis used for melting, with suitable additions of scrap, etc., for the production of ferrous castings.
Pilot or sample casting	A casting made from a pattern produced in a production die to check the accuracy of dimensions and quality of castings which will be made in quantity.
Porosity (blow-holes)	Holes in the casting due to gases trapped in the mold, reaction of molten metal with moisture in the molding sand, or imperfect fusion of chaplets with molten metal. (Surface porosity may be due to overheating of the mold or core faces, but should not be confused with sand inclusions.)
Recovery rate	Ratio of the number of saleable parts to the total number of parts manufactured, expressed as a percentage.
Refractory	Heat resistant ceramic material.
Reject rate	Ratio of the number of parts scrapped to the total number of parts manufactured, expressed as a percentage.
Riser	(See *feeder*.)
Runner system or gating	The set of channels in a mold through which molten metal is poured to fill the mold cavity. The system normally consists of a vertical section (downgate or sprue) to the point where it joins the mold cavity (gate) and leading from the mold cavity further vertical channels (risers or feeders).
Sand inclusions	Cavities or surface imperfections on a casting caused by sand washing into the mold cavity.
Scrap	(a) Any scrap metal melted (usually with suitable ad-

ditions of pig iron or ingots) to produce castings.
(b) Reject castings.

Shakeout — The process of separating the solidified casting from the mold material.

Shrinkage — Contraction of metal in the mold during solidification. The term is also used to describe the casting defect, i.e., shrinkage cavity. This results from poor design, insufficient metal feed, or inadequate feeding.

Slag — A fused nonmetallic material used to protect molten metal from the air and to extract certain impurities.

Slag inclusions — Casting surface imperfections similar to sand inclusions, but containing impurities from the charge materials, silica and clay eroded from the refractory lining, ash from the fuel during the melting process. May also originate from metal-refractory reactions occurring in the ladle during pouring of the casting.

Slurry — A flowable mixture of refractory particles suspended in a liquid. (See *dip coat*.)

Sodium silicate CO_2 process — Molding sand is mixed with sodium silicate and the mold is gassed with carbon dioxide gas to produce a hard mold or core.

Sprue (downsprue-downgate) — The channel, usually vertical, which the molten metal enters: so-called because it conducts metal down into the mold.

Test bar — Standard specimen bar designed to permit determination of mechanical properties of the metal from which it was poured.

Test lug — A lug cast as apart of the casting and later removed for testing purposes.

Vent — A small opening or passage in a mold or core to facilitate escape of gases when the mold is poured.

17
Resources

17.1 INFORMATION SOURCES

The compilation of the material included in this publication could not have been accomplished without the cooperation of the many contributors to the literature and their dedication to the advancement of the metalcasting industry. Special acknowledgment for permission to reproduce data is gratefully extended to the authors and publishers of the following publications.

Metalcasters Reference & Guide
American Foundrymen's Society, Inc.

Metalcasting & Molding Processes
American Foundrymen's Society, Inc.

No-Bake Core & Molds
American Foundrymen's Society, Inc.

Cast Metals Technology
J. G. Sylvia
Published by Addision-Wesley Pub. Co.

Guide to Buying Castings
A. H. Grant
Published by Institute of Publishing and Supply
Easton House, Easton on the Hill
Near Stamford, Lincolnshire, England

17.2 METALCASTING ASSOCIATIONS

Abrasive Engineering Society
118 Main Street / Box 441
Connoquenessing, PA 16027 / (412) 789-7596

The Aluminum Association, Inc.
900 19th Street, N.W.
Washington, DC 20006 / (202) 862-5100

American Cast Metals Association
455 State Street
Des Plaines, IL 60016 / (312) 299-9160

American Die Casting Institute, Inc.
2340 Des Plaines Avenue, Suite 307
Des Plaines, IL 60018 / (312) 298-1220

American Foundrymen's Society, Inc.
Golf & Wolf Roads
Des Plaines, IL 60016-2277 / (312) 824-0181

Cast Metals Institute of AFS
Golf & Wolf Roads
Des Plaines, IL 60016-2277 / (312) 824-0181

Casting Industry Suppliers Association
6990 Rieber Street
Worthington, OH 43085 / (614) 848-8199

Ductile Iron Society
615 Sherwood Parkway
Mountainside, NJ 07092 / (201) 232-3080

Foundry Educational Foundation
484 East Northwest Highway
Des Plaines, IL 60016-2202 / (312) 299-1776

International Lead Zinc Research Organization, Inc.
P.O. Box 12036 / 2525 Meridian Parkway
Research Triangle Park, NC 27709 / (919) 361-4647

International Magnesium Association
7927 Jones Branch Drive, Suite 400
McLean, VA 22102 / (703) 442-8888

Investment Casting Institute
8350 North Central Exp., Suite 306
Dallas, TX 75206 / (214) 368-8896

Iron Casting Research Institute
2838 Fisher Road, Suite B
Columbus, OH 43204-3574 / (614) 275-4201, 4202

Lead Industries Association
292 Madison Avenue
New York, NY 10017 / (212) 578-4750

National Association of Pattern Manufacturers
21010 Center Ridge Road
Cleveland, OH 44116 / (216) 333-7417

Non-Ferrous Founders' Society
455 State Street, Suite 100
Des Plaines, IL 60016 / (312) 299-0950

Society of Die Casting Engineers
2000 North Fifth Avenue
River Grove, IL 60171 / (312) 452/0700

Steel Founders' Society of America
Cast Metals Federation Building
455 State Street
Des Plaines, IL 60016 / (312) 299-9160

Zinc Institute, Inc.
292 Madison Avenue
New York, NY 10017 / (212) 578-4750

17.3 DIRECTORY OF SPECIFICATION SOURCES

AASHTO
American Association of State Highway & Transportation Officials
444 North Capitol
Washington, DC 20001 / (202) 624-5800

AGMA
American Gear Manufacturers Assn.
1500 King Street, Suite 201
Alexandria, VA 22314 / (703) 684-0211

AMS
Aeronautical Materials Specifications of the Society of Automotive Engineers
400 Commonwealth Drive
Warrendale, PA 15096 / (412) 776-4841

ANSI
American National Standards Institute, Inc.
1430 Broadway
New York, NY 10018 / (212) 354-3300

API
American Petroleum Institute
1220 L Street, N.W.
Washington, DC 20005 / (202) 682-8000

ASME
American Society of Mechanical Engineers
United Engineering Center
345 East Forty-Seventh Street
New York, NY 10017 / (212) 705-7722

ASTM
American Society for Testing and Materials
655 – 15th Street, N.W.
Washington, DC 20005 / (202) 639-4025

AWS
American Welding Society
550 LeJeune Road, N.W.
Miami, FL 33135 / (305) 443-9353

AWWA
American Water Works Association, Inc.
6666 West Quincy Avenue
Denver, CO 80235 / (303) 794-7711

CDA
Copper Development Association
Box 1840, Greenwich Office Park 2
Greenwich, CT 06836 / (203) 625-6658

SAE
Society of Automotive Engineers, Inc.
400 Commonwealth Drive
Warrendale, PA 15096 / (412) 776-4841

Federal

U.S. Naval Supply Depot
5801 Tabor Avenue
Philadelphia, PA 19120 / (215) 697-3321

Military
U.S. Naval Supply Depot
5801 Tabor Avenue
Philadelphia, PA 19120 / (215) 697 3321

Army, Navy, and Air Force Specifications, including Mil and Jan, may also be obtained from the service concerned—Army: Army supply installations throughout the U.S.

Navy: Commanding Officer, U.S. Naval Supply Depot, 5801 Tabor Avenue, Philadelphia, PA 19120
Air Force: Commander, USAF Engineering Specifications and Drawing Branch, Administrative Services Office, Attn: EWBFE, Wright-Patterson Air Force Base, Ohio

MSS
Manufacturer Standardization Society of the Valve and Fittings Industry
127 Park N.E.
Vienna, VA 22180 / (703) 281-6613

QQ
(Federal and Military)
General Services Administration in your area

Ingot No.
Brass and Bronze Ingot Institute
33 North LaSalle Street, Room 3500
Chicago, IL 60602 / (312) 236-2715

17.4 CONVERSIONS

	When you know:	You can find:	If you multiply by:
LENGTH	inches	millimeters	25
	feet	centimeters	30
	yards	meters	0.9
	miles	kilometers	1.6
	millimeters	inches	0.04
	centimeters	inches	0.4
	meters	yards	1.1
	kilometers	miles	0.6
AREA	square inches	square centimeters	6.5
	square feet	square meters	0.09
	square yards	square meters	0.8
	square miles	square kilometers	2.6
	acres	square hectometers (hectares)	0.4
	square centimeters	square inches	0.16
	square meters	square yards	1.2
	square kilometers	square miles	0.4
	square hectometers (hectares)	acres	2.5

MASS	ounces	grams	28
	pounds	kilograms	0.45
	short tons	megagrams (metric tons)	0.9
	grams	ounces	0.035
	kilograms	pounds	2.2
	megagrams (metric tons)	short tons	1.1
LIQUID VOLUME	ounces	milliliters	30
	pints	liters	0.47
	quarts	liters	0.95
	gallons	liters	3.8
	milliliters	ounces	0.034
	liters	pints	2.1
	liters	quarts	1.06
	liters	gallons	0.26
TEMPERATURE	degrees Fahrenheit	degrees Celsius	5/9 (after subtracting 32)
	degrees Celsius	degrees Fahrenheit	9/5 (then add 32)

VELOCITIES

Feet per second per second	Miles per hour per second	Meters per second per second	Kilometers per hour per second
1	0.6818	0.3048	1.09728
1.4667	**1**	0.44704	1.6093
3.2808	2.237	**1**	3.6
0.9113	0.6214	0.2778	**1**

POWER

Foot-pounds per second	Foot-pounds per minute	Horsepower	Watts	Kilowatts
1	60	0.001818	1.3557	0.001356
0.01667	**1**	$0.0_4 3030$	0.0226	$0.0_4 226$
550	33000	**1**	745.65	0.74565
0.7376	44.257	0.001341	**1**	0.001
737.6	44256	1.341	1000	**1**

PRESSURES

Pounds per in.^{2}	Pounds per ft.^{2}	Short tons per ft.^{2}	Atmo-spheres	Columns mercury at 0°C., inches	Column water at 15°C., feet	Column water at 15°C., inches	Kilograms per cm^2
1	144	0.0720	0.06804	2.036	2.309	27.70	0.07031
0.00694	1	0.0005	$0.0_{3}4725$	0.01414	0.01603	0.1924	$0.0_{3}4882$
13.89	2000	1	0.9450	28.28	32.06	384.8	0.9765
14.70	2116.3	1.058	1	29.92	33.93	407.2	1.0333
0.4912	70.73	0.03536	0.03342	1	1.134	13.61	0.03453
0.4332	62.43	0.03119	0.02947	0.8819	1	12	0.03045
0.03610	5.2023	0.002599	0.002456	0.07349	0.08333	1	0.002538
14.22	2048	1.024	0.9678	28.96	32.84	394.0	1

17.5 FLOW CHART OF A TYPICAL FOUNDRY OPERATION

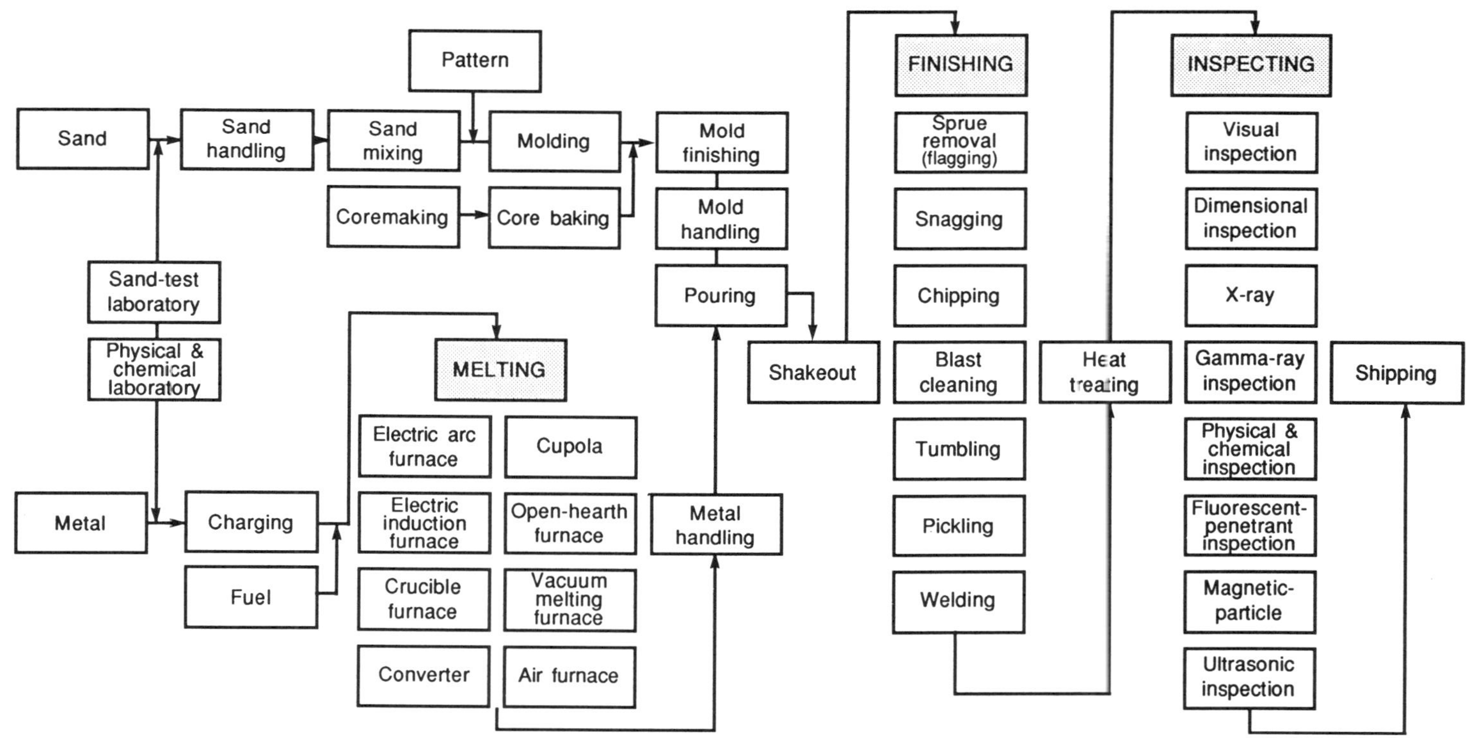

17.6 REQUEST FOR QUOTATION FORM

A. IDENTIFICATION

Company

Address

Individual Title

Pattern No. Part No.

B. QUANTITY

Total number required: If subject to release, number per release

Estimated yearly requirements of this casting pieces.

Is this a ☐ Current Production Part?

☐ New Production Part?

☐ Experimental Development Part for Test?

C. DELIVERY SCHEDULE

Initial Delivery (date) Number of Pieces

Subsequent Schedule pieces per month.

Order to be Complete by (date)

Other Delivery Comments

D. ACCOMPANYING BLUEPRINT CHECKLIST

Check the following items when shown on the blueprint.

☐ Actual casting weight or ☐ Estimated weight.

☐ Pattern Number ☐ Part No.

☐ Locating and/or clamping bosses for machining are shown.

☐ Finish stock requirements are given.

☐ Type of finishing operation to be used is given.

☐ Location for pattern numbers, trade marks, heat numbers, etc.

☐ Raised characters ☐ Sunken characters

☐ Changeable inserted numbers

E. CASTING SERVICE DATA

1. Part Name

2. Service Stresses

psi maximum design load % safety factor required

☐ Subject to mild impact ☐ Subject to heavy impact.

☐ Subject to rapid cyclic loading (endurance).

☐ Subject to hydraulic pressure or ☐ pneumatic pressure.

3. Wear or Abrasion

☐ Subject to wear against (name other material)

☐ Good lubrication ☐ Intermittent lubrication ☐ No lubrication

☐ Abrasive wear by (name material)

4. Corrosion

Type of material or environment at °F

☐ Strongly acid ☐ Strongly alkaline or pH (give value)

☐ Exposed to air or aerated liquid; ☐ Submerged and relatively air free.

☐ Media contamination is important; ☐ Not important.

5. Elevated Temperature Service

°F is maximum service temperature of casting.

☐ Steady temperature during use.

☐ Undulating temperature °F to °F.

☐ Shock heating or cooling. Describe

☐ Surface Scaling is critical.

☐ Dimensional stability is critical.

☐ Load carrying ability is critical.

Comments

6. Machining Operations To Be Performed:

☐ None.

☐ Extensive.

☐ Nominal.

☐ On conventional machine tools.

☐ By high energy grinding.

☐ On automatic machine tools.

☐ Locating points are shown on drawing.

☐ Casting to be targeted by foundry.

7. Casting Will be Otherwise Finished By:

☐ Painting only. ☐ Porcelain enamel.

☐ Filling, rubbing and painting.

☐ Plating, kind

☐ None ☐ Other, (name)

F. CASTING PROPERTIES

1. Standard Specification No. issued by

☐ Covers requirements completely. No other data are necessary.

☐ Is basis with additional requirements below.

☐ No standard specification applies.

2. Strength

Tensile strength of psi minimum using

A.S.T.M. Test Bar Size: ☐ "A" ☐ "B" ☐ "C" ☐ Other

If other, specify

Certification is required for: ☐ Chemical Analysis;

☐ Mechanical Properties.

3. Hardness

BHN maximum; BNH minimum at location marked on drawing.

4. Machinability

☐ No heat treatment required.

☐ Heat treatment required to meet specifications.

☐ Stress relief required. ☐ Softening anneal required.

☐ Hardening required (details attached as to type, location, and hardness).

☐ To be done by foundry. ☐ To be done by customer.

6. Casting Appearance

☐ Is critical. ☐ Is important. ☐ Is not important.

7. Dimensional Tolerances

☐ Commercial tolerances are satisfactory.

☐ Close tolerances apply where marked on drawing.

☐ Gaging is required as follows:

☐ Gages or special fixtures are available as follows:

G. SPECIAL INSPECTION METHODS

1. To be Performed By:

Foundry	Customer	
☐	☐	Pressure test, State requirements:
☐	☐	X-Ray Requirements:
☐	☐	Sonic Testing Requirements:
☐	☐	Magnetic Particle Testing Requirements:
☐	☐	Liquid Penetrant Testing Requirements:
☐	☐	Other Testing Requirements:

2. Customer's Method of Incoming Inspection

☐ Full inspection, rejection of only defective.

☐ Statistical inspection, rejection of lot on basis of sample.

Sampling ratio ; Rejection No.

Based on AQL (Average Quality Level) of

H. SHIPPING

☐ Usual Procedure

☐ Other packing or crating is required as follows:

☐ Motor Freight ☐ Rail ☐ Other

I. PATTERN EQUIPMENT

☐ New pattern equipment is to be furnished by foundry at customer's expense.

☐ Foundry is to specify equipment to be made by customer's pattern maker.

☐ Pattern equipment is already available.

1. Type of Pattern (Check all applicable factors)

☐ Loose

☐ Solid ☐ With Follow-Board ☐ Sweep

☐ Split ☐ Can be Mounted ☐ Skeleton

☐ Match Plate; Number of castings on plate

Other patterns on plate as follows:

☐ Cope and Drag Set ☐ Single plate for cope and drag.

Number of castings per mold

☐ Other (insert, automatic molding, etc.)

3. Pattern Material

☐ Soft Wood ☐ Hard wood ☐ Wood with metal reinforcement
☐ Aluminum ☐ Plastic ☐ Iron
☐ Other

3. Core Equipment

Number of complete cores per casting: of the same core, or different cores.

Number of complete cores per box: Number of core boxes

Boxes are made from:

☐ wood ☐ aluminum ☐ plastic

Equipment suited for:

☐ Blowing
☐ Shell process
☐ Hot box process
☐ Gas-hardened sand
☐ Self-hardening sand

Describe any core gages or jigs

J. FLASK EQUIPMENT

1. Flask Size of Mounted or Plate Patterns

Size " x " Cope height ; Drag depth

☐ None ☐ Yes, will be provided; Number Type

Size x Cope height ; Drag depth

■18

References and Recommended Reading

The references listed below are a representative selection of the great amount of existing material available. For information relative to additional data, or for assistance in procuring any of the listed references, please contact the AFS Library.

MOLDING AND CASTING PROCESSES

Kotzin, E.L., "Metalcasting and Molding Processes," An American Foundrymen's Society Publication (1981) p. 246

Morgan, A.D., "Highest Quality Casting — Which Molding Process?" Foundry Trade Journal, V 152, N 3237 (May 6, 1982); p. 611-612, 614-616, 618-619, 621-622 (9 pages)

Riley, P.H., "Casting Processes," FWP Journal, V 221, N 10 (October 1982) p. 49-50, 52, 54, 56, 60, 62, 64 (8 pages)

CONVENTIONAL MOLDING PROCESSES

American Foundrymen's Society, High-Pressure Molding, 2nd edition GM8603

Ashby, G., "High Density and Automatic Molding — Do It Right," Proceedings of the International Foundry Conference 79, Johannesburg, South Africa (February 6-9, 1979) (15 pages)

Boenisch, D., "Impact Compaction of Green Sand Molds," Foundry Management & Technology, V 11, N 7 (July 1983) p. 54-57 (4 pages)

Heine, H.J., "Green Sand Molding Technology — Review and Outlook," Foundry Management and Technology, V 114, N 4 (April 1986) p. 22-24, 26, 28, 30 (6 pages)

Jacobs, B.E., "Why Green Sand?" Green Sand Technology, Proceedings of AFS-CMI Conference, Des Plaines, Illinois (March 23-24, 1983) p. 1-3 (3 pages)

Lobenko, I.P., et al, "Methods of Compacting Foundry Molds by Squeezing," Russian Casting Production (October 1972) p. 385-388

Noble, I.A., "Impact Molding Broadens Scope for the Greensand Process," Foundry Trade Journal International, V 6, N 16 (December 1982) p. 19-20 (2 pages)

Schaum, J.G., "Green Sand Molding: From Squeezers to Automation," Australian

Foundry Institute, 17 Annual Conference, Sydney (October 17-20, 1982) Paper 5 (8 pages)

Stack, R.E., "Automation Vertically Parted High Density Green Sand Molding," Green Sand Technology, Proceedings of AFS-CMI Conference, Des Plaines, Illinois (March 23-24, 1983) p. 57-72 (16 pages)

Wootten, R., "Good Sand Management for Best Results when Using High Density Molding," Foundry Trade Journal, V 160, N 3324 (April 10, 1986) p. 278, 280, 282, 284 (4 pages)

PRECISION MOLDING AND CASTING PROCESSES

Permanent Molding

Bjoklund, D., "Advantages of the Permanent Mold Process for Aluminum Castings," ASTME Seminar Presentation, 17 p.

Clark, A.P., "An Update on Permanent Mold Castings," 1st International Conference on Austempered Ductile Iron: Your Means to Improved Performance, Productivity and Cost. ASM, Rosemont, Illinois (April 2-4, 1984) p. 215-225 (11 pages)

Clark, A.P.; Godsell, B.C., "Answers to Questions about Ferrous Permanent Molding," Modern Casting V 74, N 2 (February 1984) p. 25-28 (4 pages)

Heine, H.J., "Making Ferrous Castings in Permanent Molds — Part I," Foundry Management & Technology, V 111, N 5 (May 1983) p. 28, 30, 32 (3 pages)

Heine, H.J., "Permanent Molding Gets a New Look," Foundry Management & Technology, V 110, N 7 (July 1982) p. 60-63 (4 pages)

Irving, R.R., "Permanent Molds Improve Castings," Iron Age, V 229, N 4 (February 21, 1986) p. 41-42, 47 (3 pages)

Sedzmir, A.; Parent-Simonin, S.; Parisien, J., "Permanent Molding of Ductile Iron," Modern Casting Tech Report No. 7710

Stahl, G., "Pattern Requirements of Permanent Molding Tooling," AFS Transactions, vol 77 (1969) p. 15-16

Stahl, G.W.; Whaler, K.R., "Reviewing the Permanent Mold Process — Part II," Modern Casting, V 71, N 11 (November 1981) p. 54-58 (5 pages)

"Vacuum Permanent Mold Aluminum Castings," Automotive Engineering, V 93, N 12 (December 1985) p. 52-56 (5 pages)

Diecasting

Chatterton, J.V., "The Techniques of Pressure Diecasting," Foundry Welding and Production Engineer (June 1973) p. 23-27

Cross, R., "Ferrous Diecasting," First South Pacific Diecasting Congress, Melbourne, Australia (March 10-13, 1980) Paper 80-17, p. 121-122 (2 pages)

Jchno, A., "Low Pressure Die Casting at a Sea," First South Pacific Diecasting Congress, Melbourne, Australia (March 10-13), 1980) Paper 80-29, p. 123-127 (2 pages)

Investment Casting

Argueso, L., "History of Pattern Waxes Composition, Characteristics and Handling," Steel Founders' Research Journal, N 15 (3rd Quarter 1986) p. 15-18 (4 pages)

Barbero, R.J., "In-Depth Survey of the U.S. Casting Investment Casting Industry," British Investment Casting Trade Association 17th Annual Conference, Stratford-Upon-Avon (September 4-7, 1983) Paper 16 (10 pages)

Bidwell, H.T., "Investment Castings," The Machinery Publishing Co., Ltd., New England House, New England St., Brighton, Sussex BN41 4HN

Chandley, G.D., "Automated Counter Gravity Investment Casting," Casting Engineering & Foundry World V 15, N2 (Summer 1983) p. 51-54, 56 (5 pages

Chandley, G.D., "Stainless Steel Investment Castings," Stainless Steel Castings, ASTM Special Technical Publication 756, (1982) p. 78-91 (7 pages)

Coghill, T.L., "Investment Castings more than a Net Shape," Casting Engineering & Foundry World, V 16, N 2 (Summer 1984) p. 33-34, 36, 38-39 (5 pages)

Colton, G.J., "Influence of Vacuum Processing on Investment Castings," British Investment Casters' Technical Association 16th Annual Conference, Torquay (April 26-29, 1981) (13 pages)

Gould, G.; Baker, C.G., "Investment Casting Industry: Current and Future Market Trends," Sixth World Conference Investment Casting sponsored by BICTA, ICI, EICF, Washington, D.C., (October 10-13, 1984) Paper 17 (18 pages)

Hamilton, E., "Tooling for Lost Wax Investment Casting," Trans American Foundrymen's Society, V 93 (1985) Paper 85-155, p. 903-906 (4 pages)

Jackson, P.H., "Aluminum Alloy Investment Castings for the Aerospace Industry," Foundry Trade Journal (September 24, 1981) p. 481, 483-484, 487-488 (5 pages)

Koenig, M., "Quality Control for Investment Casting Waxes," Precision Metal (June 1980) p. 56-57 (2 pages)

Krohn, B.R., "The Art and Science of Investment Casting," Modern Casting, V 74, N 12 (December 1984) p. 22-26 (5 pages)

Liesner, C., "Production Properties and Applications of Titanium Investment Castings," British Investment Casting Trade Association, 17th Annual Conference, Stratford-Upon-Avon (September 4-7, 1983) Paper 7 (7 pages)

Pratt, D.C.; Wilinson, D.; Waudby, Dr. P.E., "Precision Casting MAR-M 246 DS Alloy," Sixth World Conference Investment Casting sponsored by BICTA, ICI, EICF, Washington, D.C., (October 10-13, 1984) Paper 10 (22 pages)

Richards, G.D., "Aluminum Alloy Investment Castings," Investment Casting in Britian: Foundry Trade Journal Supplement (1986) p. 25 (1 page)

Saunders, A., "New Waxes for Investment Casting," Proc. European Investment Casters' Federation, Palma de Mallorca (October 10-14, 1984) Paper 5 (3 pages)

Stroebel, G., "Role of Investment Casting in Modern Engineering Practices," FWP

Journal, V 23, N 11 (November 1983) p. 7-8, 10, 12, 14 (5 pages)

Sutton, W.H.; Morris J.R., "Application of High-Temperature Foam filters to Improve the Cleanliness of Investment Casting Alloys," Sixth World Conference Investment Casting sponsored by BICTA, ICI, EICF, Washington, D.C., (October 10-13, 1984) Paper 15 (19 pages)

Tefft, E.C., "Lost Wax Sculpture Foundry Design," Booklet available from the International Sculpture Center, N.W., Washington, D.C. 20007, 1050 Potomac Street, N.W., Washington D.C. 20007

Thiagarajan, M., "Recent Trends in Investment Casting," Proceedings 32nd Annual Convention, Institute of Indian Foundrymen, Madras (February 3-7, 1983) Paper 8 (5 pages)

Thorne, J.K., "Titanium Precision Investment Castings," Fifth World Conference Investment Casting, Florence (1980) Paper 6 (9 pages)

"Fundamentals of Investment Casting," Precision Metal, V 41, N 2 (February 1983) p. 15017-15020 (4 pages)

"Investment Casting Process: How It Works and What It Does," The Investment Casting Institute, 8521 Clover Meadow Drive, Dallas, Texas 75243

"New Automated Foundry Capable of Producing Large Investment Castings," Industrial Heating (March 1984) p. 16-17 (2 pages)

"Robots for Investment Casting," Foundry Trade Journal International, V 7, N 21 (March 1984) p. 43-44 (2 pages)

Ceramic Molding

Ashton, M.C.; Sharman, S.G.; Brookes, A.J., "The Replicast CS (Ceramic Shell) Process," Foundry Trade Journal, V 156, N 3283 (April 26, 1984) p. 36-329, 331-332 (6 pages)

Greenwood, R.E., "Ceramic Cores: Nucleus for Precision Internal Sections," Modern Casting, V 75, N 7 (July 1985) p. 29-31 (3 pages)

Greenwood, R.E., "Molded Ceramic Cores for the Production of Precision Cast Impellers," Fifth World Conference Investment Casting, Florence (1980) Paper 25 (10 pages)

Grote, R.E., "Replicast Ceramic Shell Molding: The Process and Its Capabilities," Trans American Foundrymen's Society, V 94 (1986) Paper 86-33

Molard, F.R.; Davidson, N., "Experience with Ceramic Foam Filtration of Aluminum Castings," Trans American Foundrymen's Society (1980) Paper 80-139, p. 594-600 (6 pages)

Sharkey, R.L.; Chandley, G.D., "CLA Casting of Non-Ferrous Alloys," Fifth World Conference Investment Casting, Florence (1980) Paper 12 (12 pages)

"Casting of Reactive Metals into Ceramic Molds," Sixth World Conference Investment Casting sponsored by BICTA, ICI, EICF, Washington, D.C. (October 10-13, 1984) Paper 4 (14 pages)

"Dowty Uses Shaw and Lost-Wax to Give the Best of Both Worlds," Foundry Trade Journal, v 59, N 3317 (November 21, 1984) p. 434, 436, 449, 441 (4 pages)

"The Replicast CS (Ceramic Shell)Process," Foundry Practice §209 (July 1984) p. 8-12 (5 pages)

CHEMICALLY BONDED SAND MOLDING PROCESSES

Binders

Carey, P.R; Winters, D.L., "Updating Resin Binder Processes-Part I February 1986 through Part IX November 1986," Foundry Management & Technology, V 114, N 6 (June 1986) p. 46-48, 50, 52 (5 pages)

Carey, P.R.; Sturtz, G.P., "Updating Resin Binder Processes — Part VI," Foundry Management & Technology, V 114, N 7 (July 1986) p. 84-88 (5 pages)

Kottke, R.H., "Microwave Curing Applications of Furan Foundry Binders," Trans American Foundrymen's Society (1981) Paper 81-60, p. 251-260 (10 pages)

Reynolds, J.A., "Advances in Shell-Molding Techniques," British Investment Casting Trade Association 17th Annual Conference, Stratford-Upon-Avon (September 4-7, 1983) Paper 10 (6 pages)

Tordoff, W.L.; Tenaglia, R.D., "Test Casting Evaluation of Chemical Binder Systems," AFS Transactions 1980, No. 80-74

Xupi, D., et al, "Bonding Mechanism of Cold-Setting Synthetic Resin Sand Mixtures," AFS International Cast Metals Journal (September 1981) p. 54-61 (8 pages)

Young, C.R., "Phenolic Urethane Cold Box — A Modern Approach to Casting Production," Trans-American Foundrymen's Society, V 90 (1982) Paper 82-110, p. 473-485

"No-Bake Cores & Molds," American Foundrymen's Society, Molding Methods and Materials Division, Cured Sand Committee, Publication No. GM 8001

Inorganic Binders

Jain, P.L.,; Panda, P.K., "Sodium Silicate Bonded Self-Hardening Sand System Using a Fluoride Hardener," British Foundryman (October 1980) p. 300-301 (2 pages)

MacDonald, R.M., "Sodium Silicate Is Back Again," AFS Transactions, No. 80-122 (1980) p. 451-456

Nicholas, K.E.L., "CO_2 — Silicate Process in Foundries," Published BCIRA (available AFS)

Thompson, R.N., "No-Bake Cores and Molds," AFS Publication No. GM8001

Organic Binders

Crowley, T.J., "Application of Microwave Energy to No-Bake Molding Systems," Casting Engineering and Foundry World (Spring 1981) p. 57-59, 61-63 (6 pages)

Heine, H.J., "No-Bake Cores and Molds: Review and Outlook Part I," Foundry Management & Technology (May 1981) p. 28-30, 32, 34 (5 pages)

Heine, H.J., "No-Bake Cores and Molds: Review and Outlook Part II," Foundry Management & Technology (June 1981) p. 141-146 (5 pages)

Juneja, J.L.; Dixit, M.P.; Mallaya, V.D., "Influence of Interface Temperature on Thermal Characteristics of Phenolic No-Bake Sands," Proc. 32nd Annual Convention, Institute of Indian Foundrymen, Madras (February 3-7, 1983) Paper 4 (8 pages)

Naro, R.L.; Hart, J.F., "Phenolic Urethane No-Bake Binders: Ten Years of Progress," AFS Transactions, No. 80-55 (1980) p. 57-66

Rutterman, J., "Sand and Its Effective Use in No-Bake Systems," Casting Engineering & Foundry World, V 16, N 4 (Winter 1985) p. 36-37, 39-42 (6 pages)

SPECIAL MOLDING AND CASTING PROCESSES

Irving, R.R., "Huge HSLA Castings Vie for Service in the North Sea," Iron, V 224, N 34 (December 7, 1981) p. 67, 69, 72 (3 pages)

Mihaichuk, W., "Introduction to Graphite Part I: Applications," Modern Casting, V 76, N 2 (February 1986) p. 36-38 (3 pages)

Reynolds, J.A., "SCRATA Replicast Process," Sixth World Conference Investment Casting sponsored by BICTA, ICI, EICF, Washington, D.C., (October 10-13, 1984) Paper 3 (7 pages)

Tamazake, Y., "Copper-Base Alloys for Plaster Mold Casting," Fifth World Conference Investment Casting, Florence (1980) Paper 3) 7 pages

Evaporative Pattern Casting

Ashton, M.C.; Sharman, S.G.; Brookes, A.J., "EPS Ceramic Shell Process Improves As-Cast Steel Quality — Part I," Foundry Management & Technology, V 112, N 4 (April 1984) p. 40, 42, 44, 48 (4 pages)

Ashton, M.C.; Sharman, S.G.; Brookes, A.J., "The Replicast FM (Full Mold) and CS (Ceramic Shell) Process," Transactions of the American Foundrymen's Society, V 92 (1984) Paper 84-22, p. 271280 (10 pages)

Benasteau, D.M., "Casting Using Polystyrene: More Economical than Pressure Diecasting," Using Nouvelle, N 3 (January Supplement 1985) Translation from French, p. 42-45

Brown, J.R., "Replicast Process," Foundry Trade Journal, V 156, N 3278 (February 2, 1984) p. 71-74 (4 pages)

Clegg, Dr. A.J., "Expanded-Polystyrene Molding — A Status Report," Foundry Trade Journal International, V 9, N 30 (June 1986) p. 51-52, 54, 56, 60-61 (7 pages)

Del Gaudio, G.; Serramoglia, G.; Caironi, G.; Tosi, G., "Aspects Concerning the Role of Coatings in the Production of Iron Castings by the 'Policast' Process," Metallurgical Science and Technology, V 3, N 3 (December 1985) p. 76-86 (11 pages)

Goria, C.A.; Gaudio, G. del; Caironi, G.; Selli, M., "Molding of Iron Castings with Evaporative Polystyrene Foam Patterns Immersed in Loose Sand: Prospects and Problems," Developments for Future Foundry Prosperity BCIRA Conference, Warrick (April 2-4, 1984) Paper 31 (18 pages)

Goria, C.A.; Serramoglia, G.; Caironi, G.; Tosi, G., "Coating Permeability: A Critical Parameter of the Evaporative Pattern Process," Trans American Foundrymen's Society, V 94 (1986) Paper 86-101 (12 pages)

Kohler, P.G., "EPC's Role in Flexible Manufacturing," Modern Casting, V 74, N 9 (September 1984) p. 31-33 (3 pages)

Madono, O.; Kobayashi, K., "Recent Developments of Molding Processes in the Japanese Foundry Industry," Transactions of the Japan Foundrymen's Society, V 3 (April 1984) p. 7-11 (5 pages)

Reynolds, J.A., "SCRATA Replicast Process," Sixth World Conference Investment Casting sponsored by BICTA, ICI, EICF, Washington, D.C., (October 10-13, 1984) Paper 3 (7 pages)

Sikora, E.J., "Evaporative Casting Using Expendable Polystyrene Patterns and Unbonded Sand Casting Techniques," AFS Transactions, vol 86 (1978) p. 65

Verster, D.M., "Practical Experience with the Replicast Full Mold Process at the Standard Brass Iron and Steel Foundries," Proceedings of the SCRATA 29th Annual Conference, Nottingham (June 20-21, 1984) Paper 4 (7 pages)

Wittmoser, A.; Baechelen, M., "Quality and Costs of Casting Produced in Full Molds Using Combustible Polystyrene Patterns," Hommes et Fonderes, N 104 (April 1986) Translation from French, p. 21-26

Woelke, G., "Expanded-Polystyrene Pattern Applications in German Jobbing Foundries," Foundry Trade Journal International, V 9, N 30 (June 1986) p. 62, 64, 68-69 (4 pages)

Vacuum ("V") Process Molding

Angelov, G.; Makedonski, A.; Dobrev. P.; Allipieve, D., "Adaptation and Development of the 'V-Process' in Manufacturing Synthetic-Resin Patterns," British Foundryman, V 78, N 5 (June 1985) p. 229-234 (6 pages)

Berndt, H., "V-Process and Its Current Application in a Carousel Molding Plant," Giesserei, V 70, N 11 (May 1983) Translation from German, p. 313-321 (26 pages)

Bishop, D.; Bose, S., "Mechanical Properties of V Process Molded Steel Castings," Transactions of the American Foundrymen's Society, V 91 (1983) Paper 82-28, p. 441-446 (6 pages)

Clegg, A.J., "The V-Process-Review and Current Status," Foundry Trade Journal, V 158, N 3307 (June 6, 1985) p. 472-474, 477, 486 (5 pages)

Englat, T.A.; Hollander. W.D., "Quality Assurance and V-Process Molding," Steel Founders' Research Journal, N 7 (3rd Quarter 1984) p. 15-21 (7 pages)

Gaindhar, J.L.; Jain, C.K.; Subbarathnamatah, K., "Prediction of Pattern Dimensions for V-Process Precision Castings through Response Surface Methodology," Trans American Foundrymen's Society, V 94 (1986) Paper 86-60

Green, G.A., "Superior Castings and Improved Environment from V Process," Casting, V 28, N 5/6 (May/June 1982) p. 30-34, 36 (6 pages)

Grote, R., "Comparison Studies of Steel Castings Produced by V-Process, Green Sand

and Cold Set Molding Methods," Transactions of the American Foundrymen's Society, V 90 (1982) Paper 82-29, p. 93-102 (10 pages)

Jain, C.K.; Gaindhar, J.L., "Simulating Behavior of Sand under Vacuum in V-Process," Institute of Indian Foundrymen, 35th Annual Convention, Calcutta (1986) p. 81-87 (7 pages)

Kasai, H., "High Production of Thin-Wall Castings by the V-Process," Trans American Foundrymen's Society (1980) Paper 80-88, p. 535-540 (6 pages)

Miura, T., "Casting Production by 'V' Process Equipment," AFS Transactions, vol 84 (1876) p. 233-236

Schneider, P., "Production of Castings by the Vacuum-Sealed Molding Process Using Binderless Sand," Foundry Trade Journal (June 6, 1974) p. 723-733

Troland, W.W., "The New V-Process is 10 Years Old," Casting Engineering & Foundry World, V 14, 1 (Spring 1982) p. 42-47 (5 pages)

Centrifugal Castings

Burden, E., "Centrifugal Casting in Zircon Molds: The Noble Process," Modern Casting, V 72, N 12 (December 1982) p. 20-22 (3 pages)

Bruni, L.; Fazzina, G., "Recent Development Aiming at Reducing Wear Rates of Cylinder Liners in Grey Iron," Order only from SAE, Dept. 695, 400 Commonwealth Drive, Warrendale, Pennsylvania 15096

Cox, A., "Centrifugal Casting in Silicon Molds," Tin and Its uses, N 148 (1986) p. 15-17 (3 pages)

Gibson, D., "Noble Process," Foundry Trade Journal, V 153, N 3247 (September 23, 1982) p. 445, 457 (2 pages)

Gibson, D.S., "Basic Design of Centrifugal Casting Machines," Foundry Trade Journal, V 156, N 3278 (February 2, 1984) p. 61-62 (2 pages)

Gonicburg, J.A.; Ritch, M.L., "Principles of Centrifugal Rubber Mold Casting"

Janco, Nathan; Centrifugal Casting, American Foundrymen's Society, Inc. (1988)

McManus, G.J., "Centrifugal Casting of High Chrome Rolls Winning Acceptance in U.S.," Iron Age, V 226, N 1 (January 3, 1983) p. 28, 30 (2 pages)

Moseley, P.J., "Developments in Iron Spun Production," British Foundrymen, Supplement (June 1979) p. 177-182 (6 pages)

Neuhauser, H.J.; Veutgen, H.J., "Bimetal Cylinder Liners by the Centrifugal Casting Process for Heavy-Duty Diesel Engines," Giesserei, V 71, N 13/14 (June 18, 1984) p. 548-558 (11 pages)

Petrik, Z., "Centrifugal Casting of Grey Iron Pipes Using the Wet Spray Method," Slevarenstvi, V 27, N 7 (1979) p. 287-289

Putchinski, J., "Centrifugally Cast Finishing Rolls," Modern Casting (January 1975), p. 46

"Stainless Steel Centrifugal Castings," Foundry Trade Journal (August 13, 1970), p. 217-219

Wallace, J.F., "Engineering Aspects of Centrifugal Castings," AFS Transactions, vol 61 (1953) p. 701

"Horizontal Centrifugal Cast Rolls at Mackintosh Hemphill," Iron and Steel Engineer, V 60, N 12 (December 1983) p. 63-64 (2 pages)

Sandusky Foundry: A Specialty Performance for Refractories," Refractory, 2(4) (1979) p. 3-7 (5 pages)

Continuous Casting

Kosarski, Z.Z.J., "The Continuous Casting Process with Special Reference to Cast Iron," The British Foundryman (August 1977) p. 235-240

Krall, H.; Douglas, B.R., "The Economic Production of Shapes and Tubes In Gray and Alloyed Iron By Horizontal Continuous Casting," AFS Transactions, vol 79 (1971) p. 31-36

Replica Molding/Spincasting

Gonicburg, J.A.; Ritch, M.L., "Principles of Centrifugal Rubber Mold Casting," available from : The Oster Group, Providence, Rhode Island 02909

Schaer, L.S., "Rubber Mold Spin Casting for Low Cost Prototype or Production Parts," Diecasting Bulletin (Australia), N 47 (August 1983) p. 10-12 (3 pages)

Horizontal ("H") Casting Process

Hoult, F.E., "The 'H' Process of Repetition Casting Manufacture," AFS Transactions, vol 87 (1979) p. 237-239

Jenkins, L.; Eberhardt, C.M., "Up-date of the 'H' Process," Modern Casting Tech Report No. 8005

19
Pictorial Dictionary

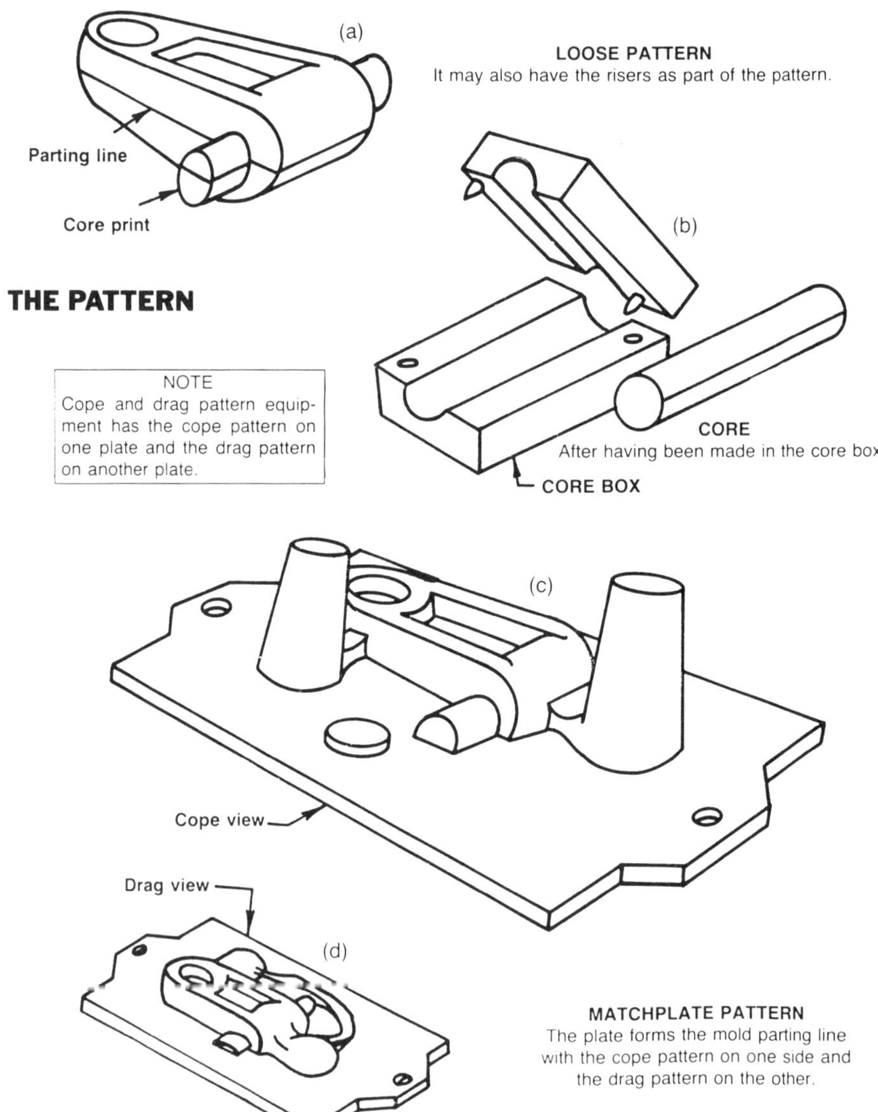

THE MOLD

OPEN MOLD SHOWING RELATION OF PARTS

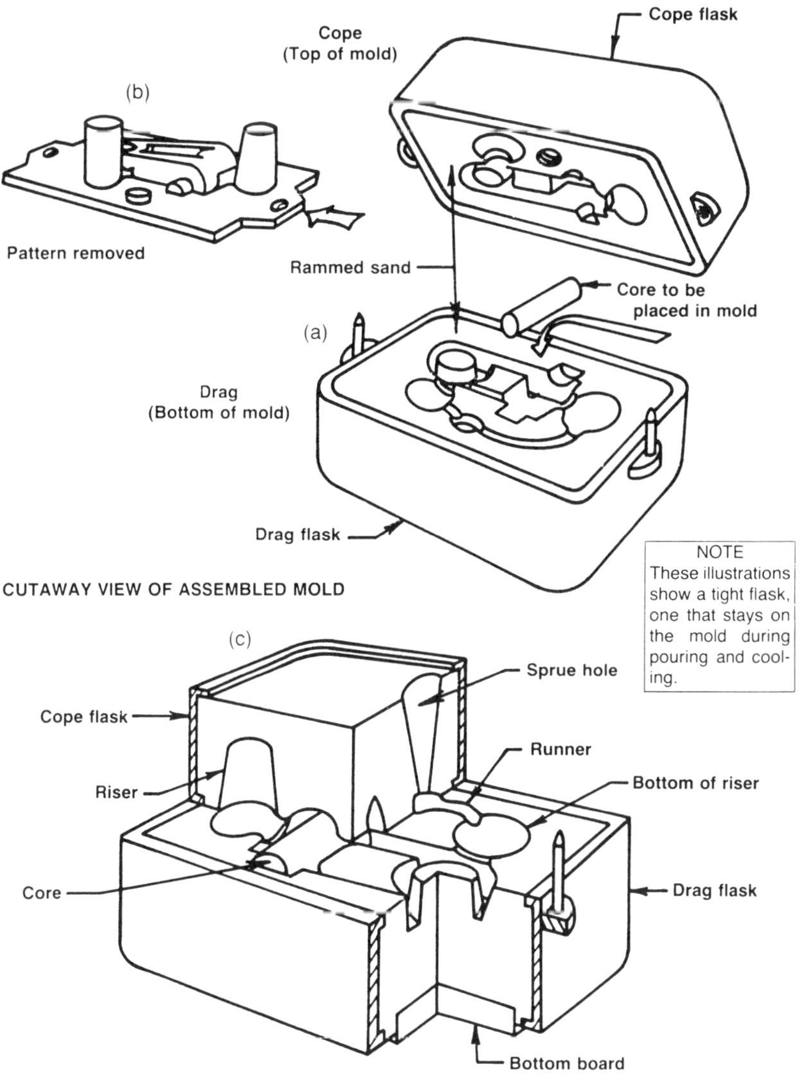

THE CASTING

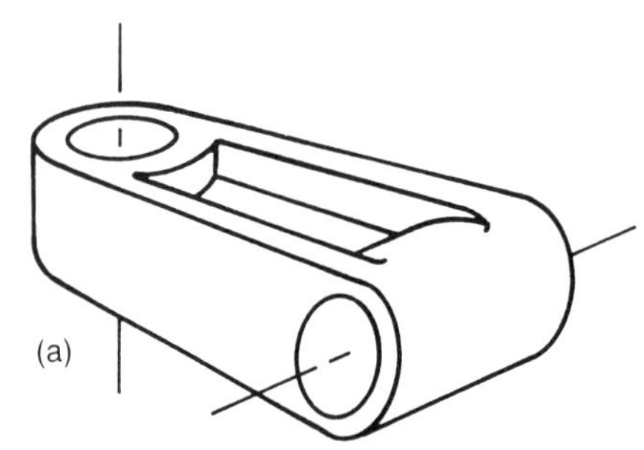

THE NEEDED FINAL CASTING

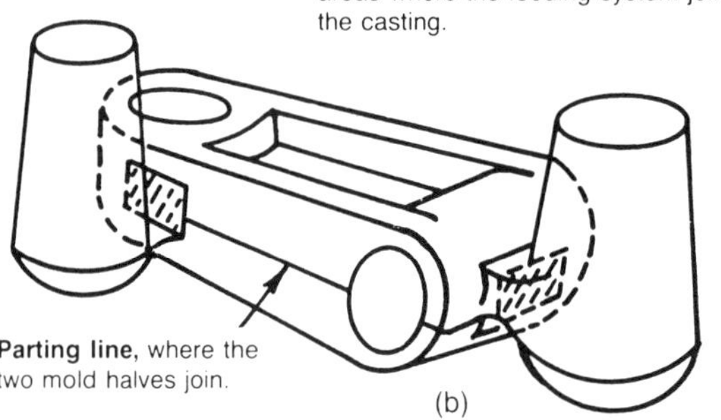

CASTING WITH FEEDERS ADDED
(Also called risers or bobs.) These will provide the reservoir of liquid metal to keep the casting cavity filled as the metal contracts during solidification.

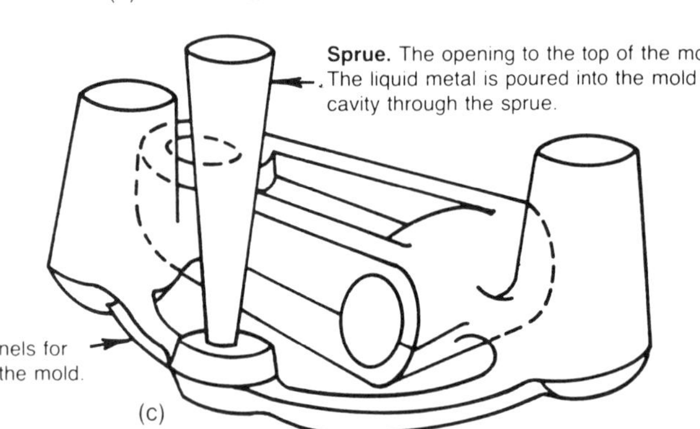

CASTING WITH COMPLETE GATING AND FEEDING SYSTEM

CORELESS INDUCTION FURNACE

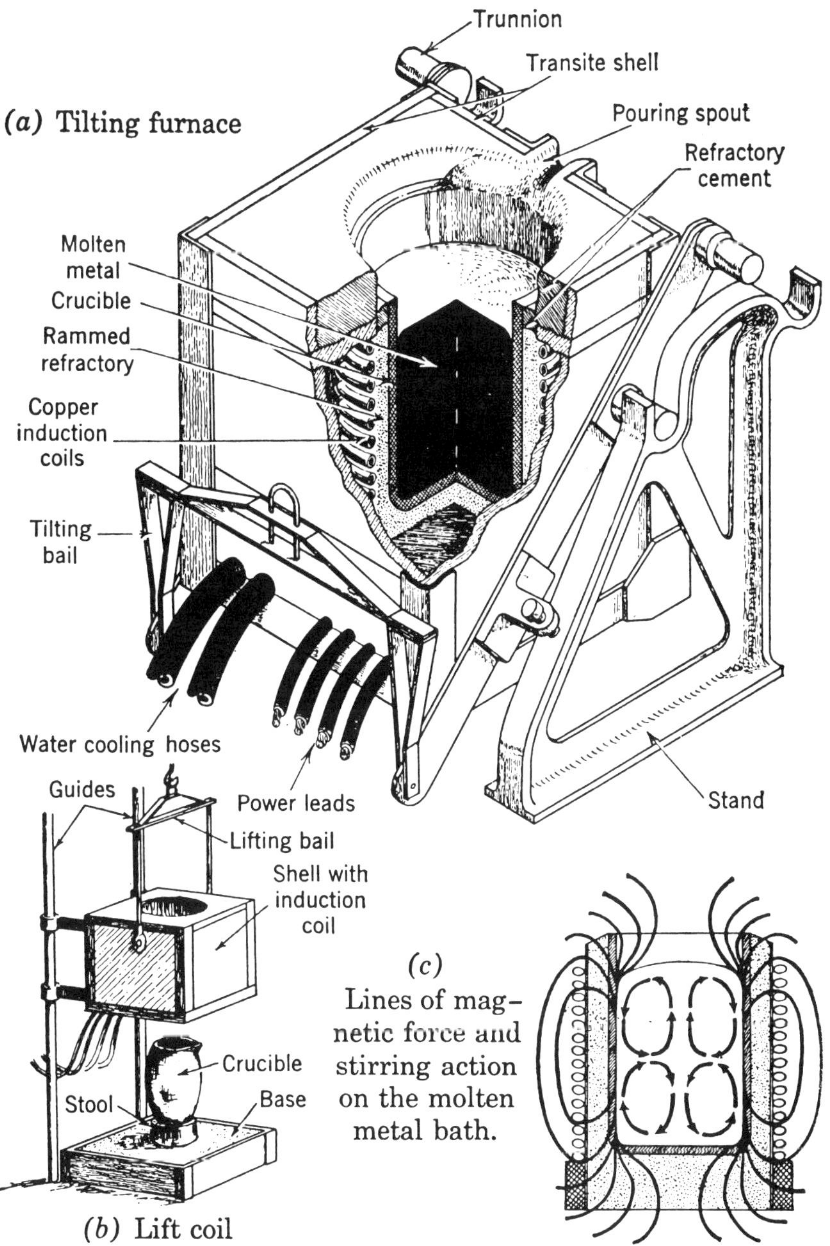

(*a*) Tilting furnace

(*b*) Lift coil

(*c*) Lines of magnetic force and stirring action on the molten metal bath.

CHANNEL INDUCTION FURNACE

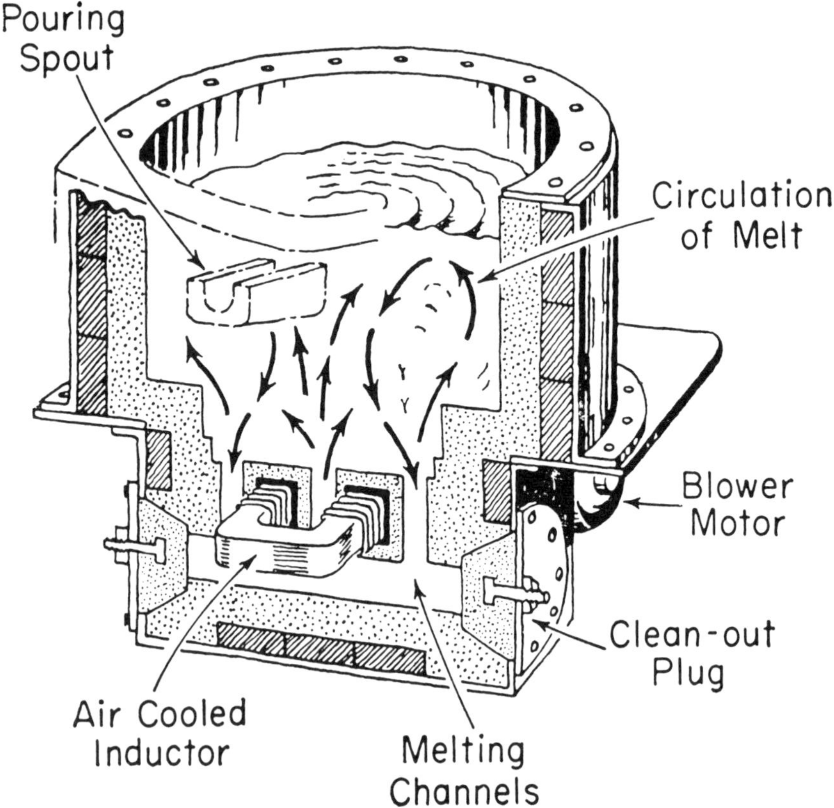

ARC FURNACES

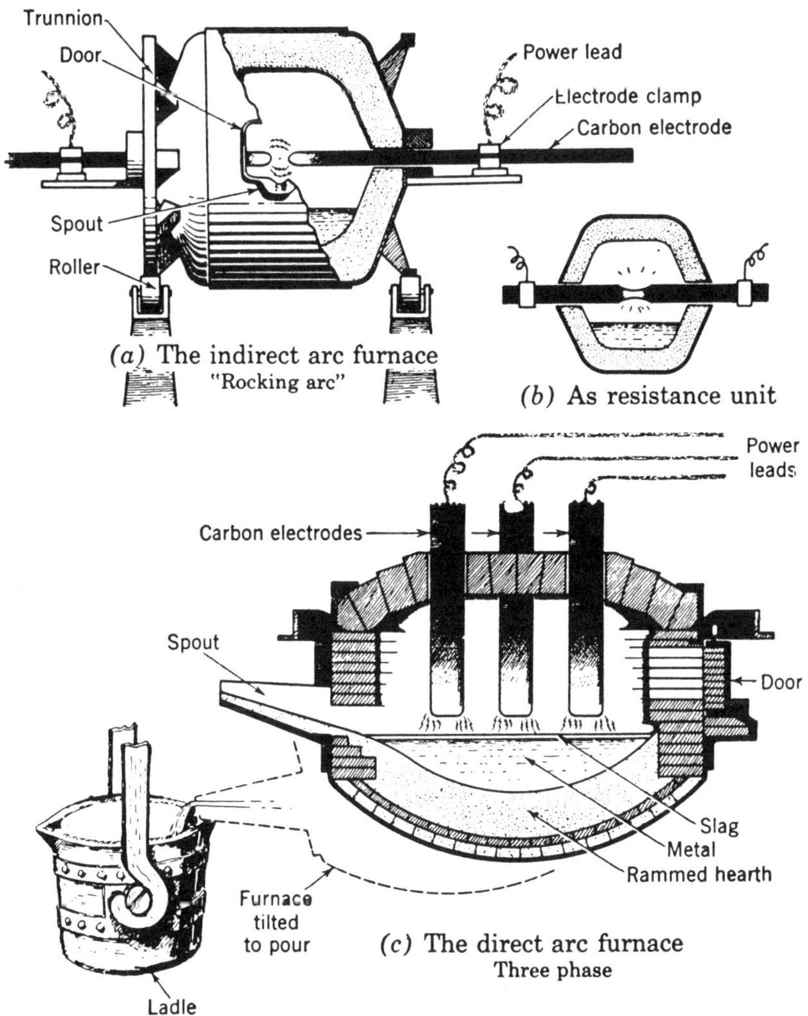

(a) The indirect arc furnace
"Rocking arc"

(b) As resistance unit

(c) The direct arc furnace
Three phase

OPEN HEARTH FURNACE

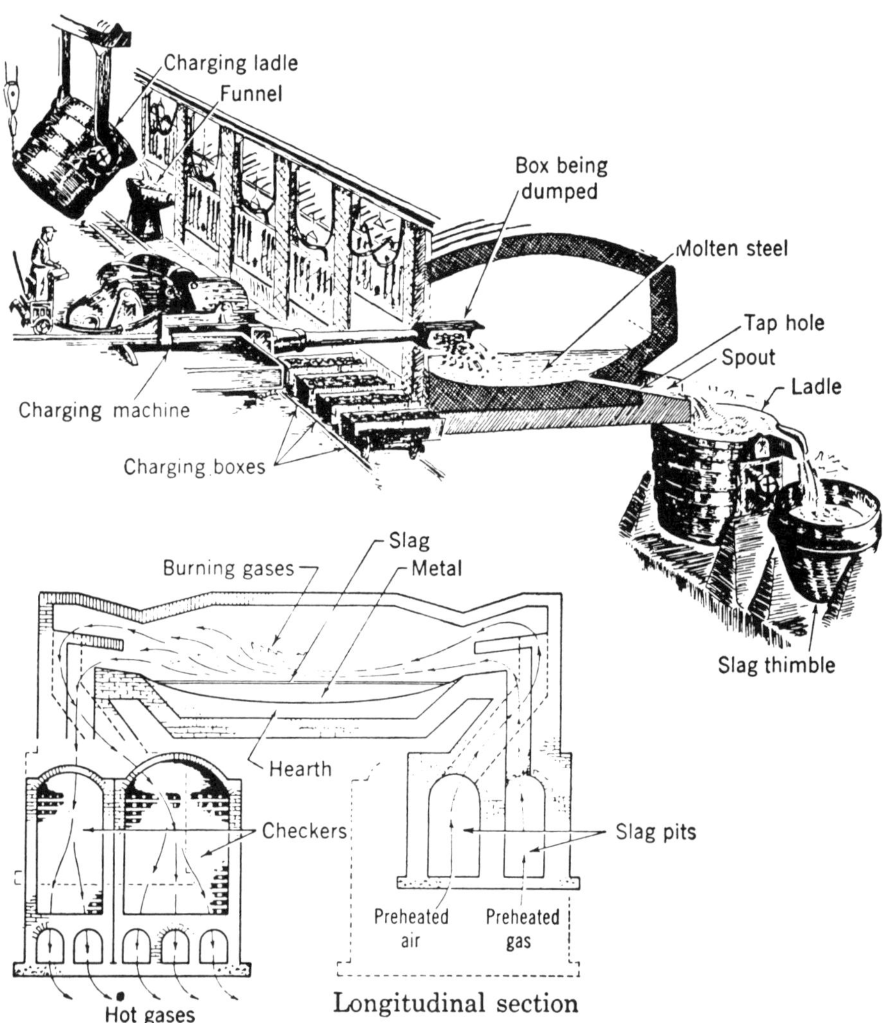

Longitudinal section

AIR FURNACE

Removable
Firebrick Arch
Roof Bung
Burners
Metal Bath
Sand Hearth
Spout
Firebrick End,
Sidewalls, and
Subhearth
Blower Pit

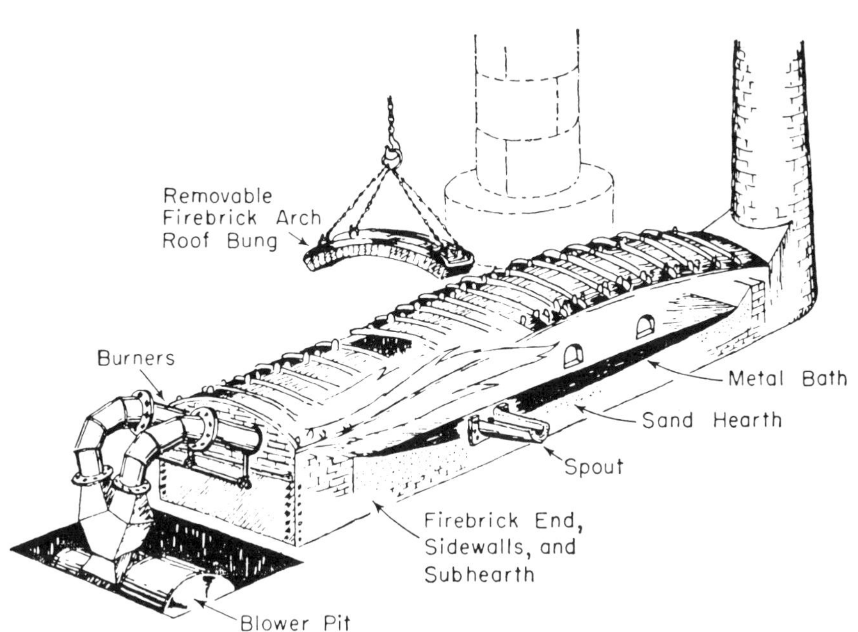

CRUCIBLE FURNACES

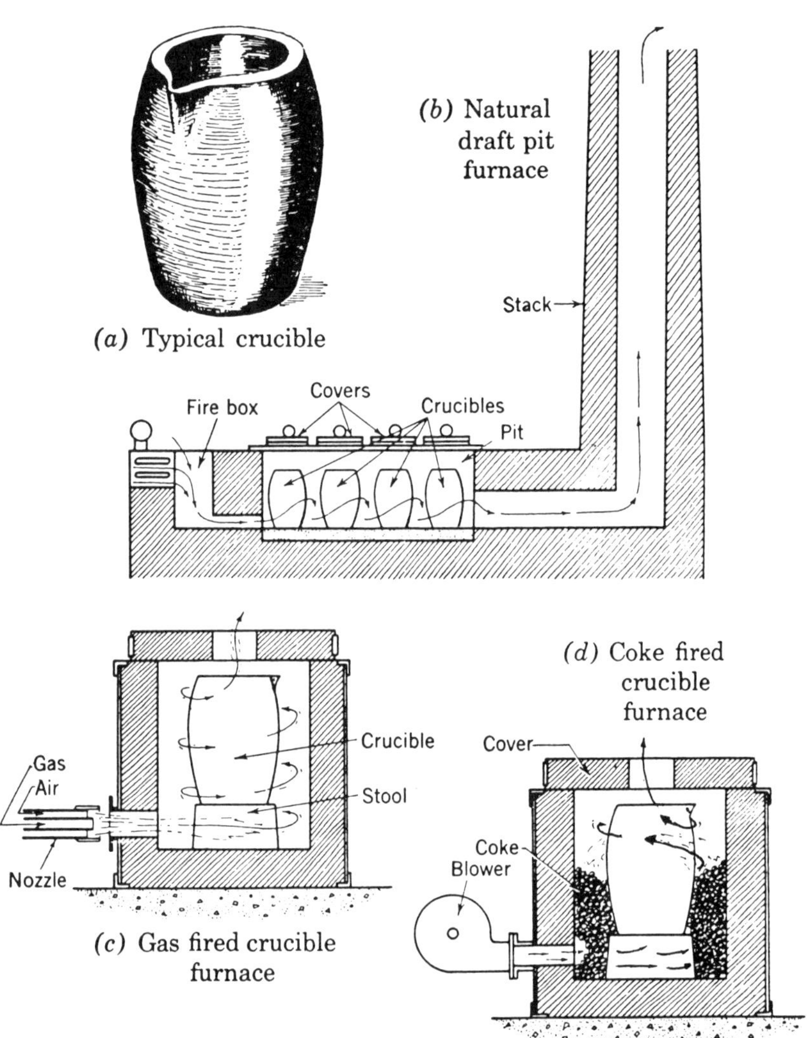

(a) Typical crucible

(b) Natural draft pit furnace

(c) Gas fired crucible furnace

(d) Coke fired crucible furnace

THE CUPOLA

Spark arrester
Shell
Lining
Stack
Charging door
Charging floor
Preheating zone
Metal charges
Limestone charges
Coke charges
Melting zone
Blast pipe
Superheating zone
Wind box
Tuyere
COKE BED
Tuyeres
SLAG
Breast
Bott
Slag spout
MOLTEN METAL
Well
Rammed sand bottom
Tapping spout
Bottom door in dropped position
Bottom prop
Foundation

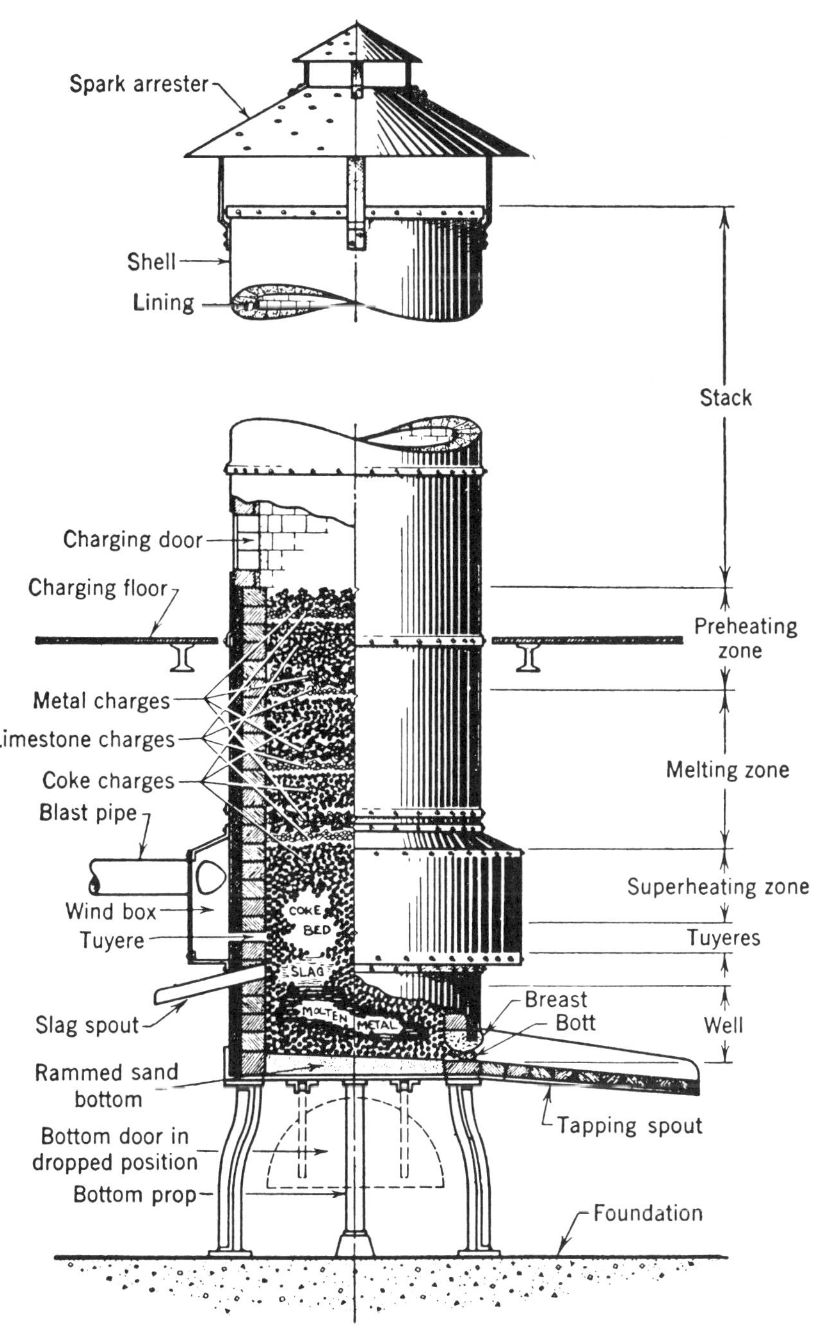

SECTIONS AND BEAMS

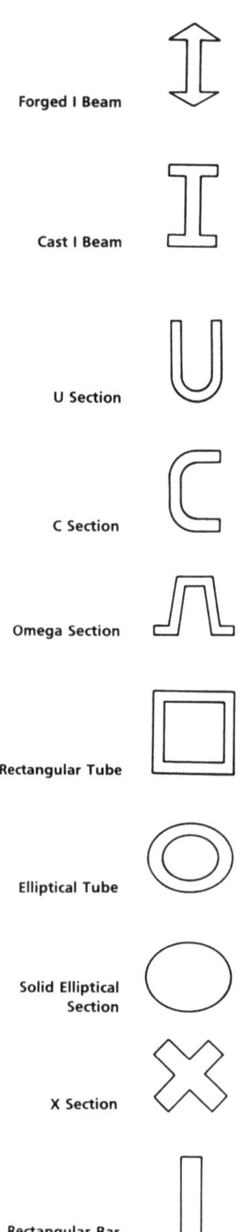

1. ***Forged I Beam.*** This is a typical forged I beam with standard forging draft. It has been selected as a basis for comparison with the cast designs, assuming a vertical load stress of 10,000 psi and a horizontal load stress also of 10,000 psi.

2. ***Cast I Beam.*** This is the cast equivalent of the forged I beam. Essentially, the only difference is the reduced draft angle and the reduced corner radii. Note the improved efficiency due to the slightly greater mass at the top and bottom. From a casting standpoint, it must be remembered that the T junction is susceptable to shrinkage, unless the junction serves as a feed channel.

3. ***U Section.*** This is a cast U section, thus avoiding the T junctions. The particular proportions for this design result in a deep pocket that may require coring. This is not a particularly efficient beam from a stress standpoint, but there are many situations where it would be a practical solution.

4. ***C Section.*** The cast C section is efficient from a stress standpoint and is an excellent foundry shape. Its primary disadvantage is that torsional stresses result when it is loaded as a beam and this disadvantage must be evaluated.

5. ***Omega Section.*** This has been called an omega section because it resembles the Greek letter omega (Ω). This is an excellent foundry shape with its uniform wall and it is an effective stress design where variable loadings are involved. Its shape is the equivalent to one cycle of a corrugated sheet.

6. ***Rectangular Tube.*** The rectangular tube is the most efficient from a stress standpoint. Loadings may be either vertical, horizontal, or torsional. The uniform section makes it a good foundry design, except that a core is required; this will involve disruption of the wall by openings for core prints, or the extensive use of chaplets.

7. ***Elliptical Tube.*** The elliptical tube is good for torsion, but it is relatively inefficient for bending loads. It, too, requires a core. Not recommended in usual circumstances.

8. ***Solid Elliptical Section.*** The solid ellipse is a good foundry shape as well as a good forging shape, but it is a poor stress section. Not recommended, except for short transition members.

9. ***X Section.*** The X section has nothing to offer. The X junction makes a poor foundry design and it is a poorly stressed beam design. Not recommended.

10. ***Rectangular Bar.*** The solid rectangular bar is a poor beam member for bending stresses.

Index

M

N

O

P

Q

R

S

T

W

Z